MONOGRAPHS ON
STATISTICS AND APPLIED PROBABILITY

General Editors

D.R. Cox, D.V. Hinkley, N. Reid, D.B. Rubin and B.W. Silverman

(Full details concerning this series are available from the Publishers.)

Density Estimation
for
Statistics and Data Analysis

B. W. SILVERMAN

School of Mathematics
University of Bath, UK

CHAPMAN AND HALL

LONDON • NEW YORK • TOKYO • MELBOURNE • MADRAS

Published by Chapman & Hall, 2-6 Boundary Row, London SE1 8HN

Chapman & Hall, 2-6 Boundary Row, London SE1 8HN, UK

Blackie Academic & Professional, Wester Cleddens Road, Bishopbriggs, Glasgow G64 2NZ, UK

Chapman & Hall, 29 West 35th Street, New York NY10001, USA

Chapman & Hall Japan, Thomson Publishing Japan, Hirakawacho Nemoto Building, 6F, 1-7-11 Hirakawa-cho, Chiyoda-ku, Tokyo 102, Japan

Chapman & Hall Australia, Thomas Nelson Australia, 102 Dodds Street, South Melbourne, Victoria 3205, Australia

Chapman & Hall India, R. Seshadri, 32 Second Main Road, CIT East, Madras 600 035, India

First edition 1986
Reprinted 1990, 1992, 1993

© 1986 B.W. Silverman

Typeset in 10/12pt Times by Thomson Press (India) Ltd., New Delhi
Printed in Great Britain by St Edmundsbury Press Ltd, Bury St Edmunds, Suffolk

ISBN 0 412 24620 1

A catalogue record for this book is available from the British Library
Library of Congress Cataloging-in-Publication Data available

Contents

Preface

The recent surge of interest in the more technical aspects of density estimation has brought the subject into public view but has sadly created the impression, in some quarters, that density estimates are only of theoretical importance. I have aimed in this book to make the subject accessible to a wider audience of statisticians and others, hoping to encourage broader practical application and more relevant theoretical research. With these objects in mind, I have tried to concentrate on topics of methodological interest. Specialists in the field will, I am sure, notice omissions of the kind that are inevitable in a personal account; I hope in spite of these that they will find the treatment of the subject interesting and stimulating.

I would like to thank David Kendall for first kindling my interest in density estimation. For their useful suggestions and other help I am grateful to several people, including Adrian Baddeley, Christopher Chatfield, David Cox, Rowena Fowler and Anthony Robinson. Colleagues who have helped in various specific ways are acknowledged in the course of the text. I am also grateful to the University of Washington (Seattle) for the opportunity to present a course of lectures that formed the basis of the book.

Bernard Silverman
Bath, May 1985

CHAPTER 1

Introduction

1.1 What is density estimation?

The *probability density function* is a fundamental concept in statistics. Consider any random quantity X that has probability density function f. Specifying the function f gives a natural description of the distribution of X, and allows probabilities associated with X to be found from the relation

$$P(a < X < b) = \int_a^b f(x)\,dx \qquad \text{for all } a < b.$$

Suppose, now, that we have a set of observed data points assumed to be a sample from an unknown probability density function. *Density estimation*, as discussed in this book, is the construction of an estimate of the density function from the observed data. The two main aims of the book are to explain how to estimate a density from a given data set and to explore how density estimates can be used, both in their own right and as an ingredient of other statistical procedures.

One approach to density estimation is *parametric*. Assume that the data are drawn from one of a known parametric family of distributions, for example the normal distribution with mean μ and variance σ^2. The density f underlying the data could then be estimated by finding estimates of μ and σ^2 from the data and substituting these estimates into the formula for the normal density. In this book we shall not be considering parametric estimates of this kind; the approach will be more *nonparametric* in that less rigid assumptions will be made about the distribution of the observed data. Although it will be assumed that the distribution has a probability density f, the data will be allowed to speak for themselves in determining the estimate of f more than would be the case if f were constrained to fall in a given parametric family.

Density estimates of the kind discussed in this book were first proposed by Fix and Hodges (1951) as a way of freeing discriminant analysis from rigid distributional assumptions. Since then, density estimation and related ideas have been used in a variety of contexts, some of which, including discriminant analysis, will be discussed in the final chapter of this book. The earlier chapters are mostly concerned with the question of how density estimates are constructed. In order to give a rapid feel for the idea and scope of density estimation, one of the most important applications, to the exploration and presentation of data, will be introduced in the next section and elaborated further by additional examples throughout the book. It must be stressed, however, that these valuable exploratory purposes are by no means the only setting in which density estimates can be used.

1.2 Density estimates in the exploration and presentation of data

A very natural use of density estimates is in the informal investigation of the properties of a given set of data. Density estimates can give valuable indication of such features as skewness and multimodality in the data. In some cases they will yield conclusions that may then be regarded as self-evidently true, while in others all they will do is to point the way to further analysis and/or data collection.

An example is given in Fig. 1.1. The curves shown in this figure were constructed by Emery and Carpenter (1974) in the course of a study of sudden infant death syndrome (also called 'cot death' or 'crib death'). The curve A is constructed from a particular observation, the degranulated mast cell count, made on each of 95 infants who died suddenly and apparently unaccountably, while the cases used to construct curve B were a control sample of 76 infants who died of known causes that would not affect the degranulated mast cell count. The investigators concluded tentatively from the density estimates that the density underlying the sudden infant death cases might be a mixture of the control density with a smaller proportion of a contaminating density of higher mean. Thus it appeared that in a minority (perhaps a quarter to a third) of the sudden deaths, the degranulated mast cell count was exceptionally high. In this example the conclusions could only be regarded as a cue for further clinical investigation.

Another example is given in Fig. 1.2. The data from which this figure was constructed were collected in an engineering experiment

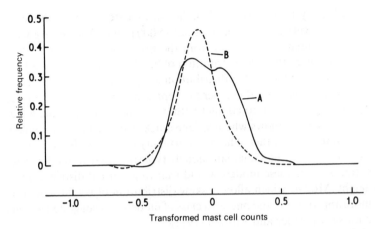

Fig. 1.1 *Density estimates constructed from transformed and corrected degranulated mast cell counts observed in a cot death study. (A, Unexpected deaths; B, Hospital deaths.) After Emery and Carpenter (1974) with the permission of the Canadian Foundation for the Study of Infant Deaths. This version reproduced from Silverman (1981a) with the permission of John Wiley & Sons Ltd.*

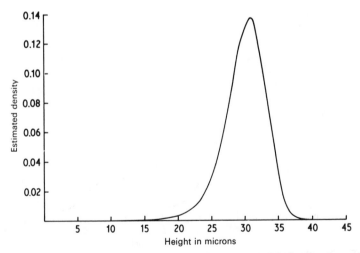

Fig. 1.2 *Density estimate constructed from observations of the height of a steel surface. After Silverman (1980) with the permission of Academic Press, Inc. This version reproduced from Silverman (1981a) with the permission of John Wiley & Sons Ltd.*

described by Bowyer (1980). The height of a steel surface above an arbitrary level was observed at about 15 000 points. The figure gives a density estimate constructed from the observed heights. It is clear from the figure that the distribution of height is skew and has a long lower tail. The tails of the distribution are particularly important to the engineer, because the upper tail represents the part of the surface which might come into contact with other surfaces, while the lower tail represents hollows where fatigue cracks can start and also where lubricant might gather. The non-normality of the density in Fig. 1.2 casts doubt on the Gaussian models typically used to model these surfaces, since these models would lead to a normal distribution of height. Models which allow a skew distribution of height would be more appropriate, and one such class of models was suggested for this data set by Adler and Firman (1981).

A third example is given in Fig. 1.3. The data used to construct this curve are a standard directional data set and consist of the directions in which each of 76 turtles was observed to swim when released. It is clear that most of the turtles show a preference for swimming approximately in the 60° direction, while a small proportion prefer exactly the opposite direction. Although further statistical model-

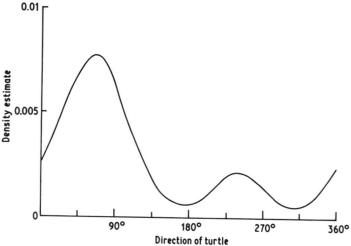

Fig. 1.3 *Density estimate constructed from turtle data. After Silverman (1978a) with the permission of the Biometrika Trustees. This version reproduced from Silverman (1981a) with the permission of John Wiley & Sons Ltd.*

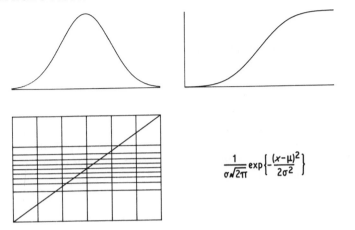

Fig. 1.4 *Four ways of explaining the normal distribution: a graph of the density function; a graph of the cumulative distribution function; a straight line on probability paper; the formula for the density function.*

ling of these data is possible (see Mardia, 1972) the density estimate really gives all the useful conclusions to be drawn from the data set.

An important aspect of statistics, often neglected nowadays, is the presentation of data back to the client in order to provide explanation and illustration of conclusions that may possibly have been obtained by other means. Density estimates are ideal for this purpose, for the simple reason that they are fairly easily comprehensible to non-mathematicians. Even those statisticians who are sceptical about estimating densities would no doubt explain a normal distribution by drawing a bell-shaped curve rather than by one of the other methods illustrated in Fig. 1.4. In all the examples given in this section, the density estimates are as valuable for explaining conclusions as for drawing these conclusions in the first place. More examples illustrating the use of density estimates for exploratory and presentational purposes, including the important case of bivariate data, will be given in later chapters.

1.3 Further reading

There is a vast literature on density estimation, much of it concerned with asymptotic results not covered in any detail in this book.

Prakasa Rao's (1983) book offers a comprehensive treatment of the theoretical aspects of the subject. Journal papers providing surveys and bibliography include Rosenblatt (1971), Fryer (1977), Wertz and Schneider (1979), and Bean and Tsokos (1980). Tapia and Thompson (1978) give an interesting perspective paying particular attention to their own version of the penalized likelihood approach described in Sections 2.8 and 5.4 below. A thorough treatment, rather technical in nature, of a particular question and its ramifications is given by Devroye and Györfi (1985). Other texts on the subject are Wertz (1978) and Delecroix (1983). Further references relevant to specific topics will be given, as they arise, later in this book.

CHAPTER 2

Survey of existing methods

2.1 Introduction

In this chapter a brief summary is given of the main methods available for univariate density estimation. Some of the methods will be discussed in greater detail in later chapters, but it is helpful to have a general view of the subject before examining any particular method in detail. Many of the important applications of density estimation are to multivariate data, but since all the multivariate methods are generalizations of univariate methods, it is worth getting a feel for the univariate case first.

Two data sets will be used to help illustrate some of the methods. The first comprises the lengths of 86 spells of psychiatric treatment undergone by patients used as controls in a study of suicide risks reported by Copas and Fryer (1980). The data are given in Table 2.1. The second data set, observations of eruptions of Old Faithful geyser in Yellowstone National Park, USA, is taken from Weisberg (1980), and is reproduced in Table 2.2. I am most grateful to John Copas and to Sanford Weisberg for making these data sets available to me.

It is convenient to define some standard notation. Except where otherwise stated, it will be assumed that we are given a sample of n real observations $X_1, \ldots, X_n$ whose underlying density is to be estimated. The symbol $\hat{f}$ will be used to denote whatever density estimator is currently being considered.

2.2 Histograms

The oldest and most widely used density estimator is the histogram. Given an *origin* x_0 and a *bin width* h, we define the *bins* of the histogram to be the intervals $[x_0 + mh, x_0 + (m+1)h)$ for positive and negative integers m. The intervals have been chosen closed on the left and open on the right for definiteness.

Table 2.1 *Lengths of treatment spells (in days) of control patients in suicide study.*

1	25	40	83	123	256
1	27	49	84	126	257
1	27	49	84	129	311
5	30	54	84	134	314
7	30	56	90	144	322
8	31	56	91	147	369
8	31	62	92	153	415
13	32	63	93	163	573
14	34	65	93	167	609
14	35	65	103	175	640
17	36	67	103	228	737
18	37	75	111	231	
21	38	76	112	235	
21	39	79	119	242	
22	39	82	122	256	

Table 2.2 *Eruption lengths (in minutes) of 107 eruptions of Old Faithful geyser.*

4.37	3.87	4.00	4.03	3.50	4.08	2.25
4.70	1.73	4.93	1.73	4.62	3.43	4.25
1.68	3.92	3.68	3.10	4.03	1.77	4.08
1.75	3.20	1.85	4.62	1.97	4.50	3.92
4.35	2.33	3.83	1.88	4.60	1.80	4.73
1.77	4.57	1.85	3.52	4.00	3.70	3.72
4.25	3.58	3.80	3.77	3.75	2.50	4.50
4.10	3.70	3.80	3.43	4.00	2.27	4.40
4.05	4.25	3.33	2.00	4.33	2.93	4.58
1.90	3.58	3.73	3.73	1.82	4.63	3.50
4.00	3.67	1.67	4.60	1.67	4.00	1.80
4.42	1.90	4.63	2.93	3.50	1.97	4.28
1.83	4.13	1.83	4.65	4.20	3.93	4.33
1.83	4.53	2.03	4.18	4.43	4.07	4.13
3.95	4.10	2.72	4.58	1.90	4.50	1.95
4.83	4.12					

The histogram is then defined by

$$\hat{f}(x) = \frac{1}{nh} \text{ (no. of } X_i \text{ in same bin as } x).$$

Note that, to construct the histogram, we have to choose both an origin and a bin width; it is the choice of bin width which, primarily, controls the amount of smoothing inherent in the procedure.

The histogram can be generalized by allowing the bin widths to

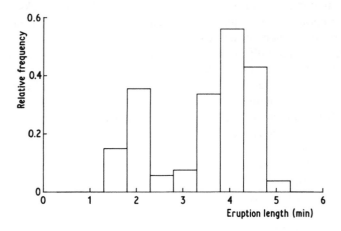

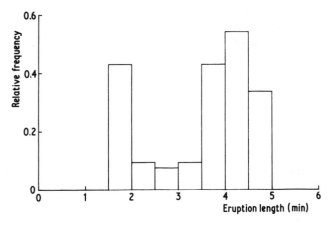

Fig. 2.1 *Histograms of eruption lengths of Old Faithful geyser.*

vary. Formally, suppose that we have any dissection of the real line into bins; then the estimate will be defined by

$$\hat{f}(x) = \frac{1}{n} \times \frac{\text{(no. of } X_i \text{ in same bin as } x\text{)}}{\text{(width of bin containing } x\text{)}}.$$

The dissection into bins can either be carried out *a priori* or else in some way which depends on the observations themselves.

Those who are sceptical about density estimation often ask why it is ever necessary to use methods more sophisticated than the simple histogram. The case for such methods and the drawbacks of the histogram depend quite substantially on the context. In terms of various mathematical descriptions of accuracy, the histogram can be quite substantially improved upon, and this mathematical drawback translates itself into inefficient use of the data if histograms are used as density estimates in procedures like cluster analysis and nonparametric discriminant analysis. The discontinuity of histograms causes extreme difficulty if derivatives of the estimates are required. When density estimates are needed as intermediate components of other methods, the case for using alternatives to histograms is quite strong.

For the presentation and exploration of data, histograms are of course an extremely useful class of density estimates, particularly in the univariate case. However, even in one dimension, the choice of origin can have quite an effect. Figure 2.1 shows histograms of the Old Faithful eruption lengths constructed with the same bin width but different origins. Though the general message is the same in both cases, a non-statistician, particularly, might well get different impressions of, for example, the width of the left-hand peak and the separation of the two modes. Another example is given in Fig. 2.2; leaving aside the differences near the origin, one estimate suggests some structure near 250 which is completely obscured in the other. An experienced statistician would probably dismiss this as random error, but it is unfortunate that the occurrence or absence of this secondary peak in the presentation of the data is a consequence of the choice of origin, not of any choice of degree of smoothing or of treatment of the tails of the sample.

Histograms for the graphical presentation of bivariate or trivariate data present several difficulties; for example, one cannot easily draw contour diagrams to represent the data, and the problems raised in the univariate case are exacerbated by the dependence of the estimates on the choice not only of an origin but also of the coordinate

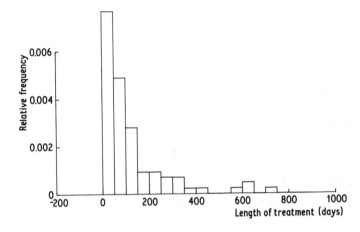

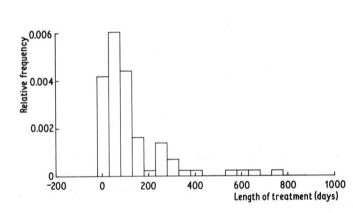

Fig. 2.2 *Histograms of lengths of treatment of control patients in suicide study.*

direction(s) of the grid of cells. Finally, it should be stressed that, in all cases, the histogram still requires a choice of the amount of smoothing.

Though the histogram remains an excellent tool for data presentation, it is worth at least considering the various alternative density estimates that are available.

2.3 The naive estimator

From the definition of a probability density, if the random variable X

has density f, then

$$f(x) = \lim_{h \to 0} \frac{1}{2h} P(x - h < X < x + h).$$

For any given h, we can of course estimate $P(x - h < X < x + h)$ by the proportion of the sample falling in the interval $(x - h, x + h)$. Thus a natural estimator $\hat{f}$ of the density is given by choosing a small number h and setting

$$\hat{f}(x) = \frac{1}{2hn} \text{ [no. of } X_1, \ldots, X_n \text{ falling in } (x - h, x + h)];$$

we shall call this the naive estimator.

To express the estimator more transparently, define the weight function w by

$$w(x) = \begin{cases} \dfrac{1}{2} & \text{if } |x| < 1 \\[2mm] 0 & \text{otherwise.} \end{cases} \tag{2.1}$$

Then it is easy to see that the naive estimator can be written

$$\hat{f}(x) = \frac{1}{n} \sum_{i=1}^{n} \frac{1}{h} w\left(\frac{x - X_i}{h}\right).$$

It follows from (2.1) that the estimate is constructed by placing a 'box' of width $2h$ and height $(2nh)^{-1}$ on each observation and then summing to obtain the estimate. We shall return to this interpretation below, but it is instructive first to consider a connection with histograms.

Consider the histogram constructed from the data using bins of width $2h$. Assume that no observations lie exactly at the edge of a bin. If x happens to be at the centre of one of the histogram bins, it follows at once from (2.1) that the naive estimate $\hat{f}(x)$ will be exactly the ordinate of the histogram at x. Thus the naive estimate can be seen to be an attempt to construct a histogram where every point is the centre of a sampling interval, thus freeing the histogram from a particular choice of bin positions. The choice of bin width still remains and is governed by the parameter h, which controls the amount by which the data are smoothed to produce the estimate.

The naive estimator is not wholly satisfactory from the point of view of using density estimates for presentation. It follows from the

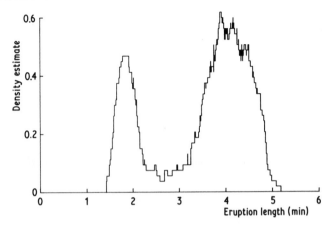

Fig. 2.3 *Naive estimate constructed from Old Faithful geyser data, h = 0.25.*

definition that $\hat{f}$ is not a continuous function, but has jumps at the points $X_i \pm h$ and has zero derivative everywhere else. This gives the estimates a somewhat ragged character which is not only aesthetically undesirable, but, more seriously, could provide the untrained observer with a misleading impression. Partly to overcome this difficulty, and partly for other technical reasons given later, it is of interest to consider the generalization of the naive estimator given in the following section.

A density estimated using the naive estimator is given in Fig. 2.3. The 'stepwise' nature of the estimate is clear. The boxes used to construct the estimate have the same width as the histogram bins in Fig. 2.1.

2.4 The kernel estimator

It is easy to generalize the naive estimator to overcome some of the difficulties discussed above. Replace the weight function w by a *kernel function K* which satisfies the condition

$$\int_{-\infty}^{\infty} K(x)\,dx = 1. \qquad (2.2)$$

Usually, but not always, K will be a symmetric probability density function, the normal density, for instance, or the weight function w

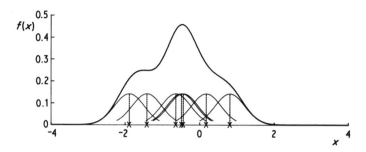

Fig. 2.4 *Kernel estimate showing individual kernels. Window width 0.4.*

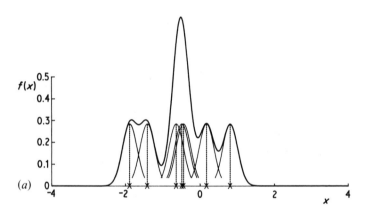

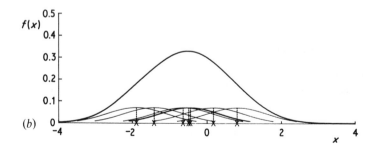

Fig. 2.5 *Kernel estimates showing individual kernels. Window widths : (a) 0.2 ;*
(b) 0.8.

used in the definition of the naive estimator. By analogy with the definition of the naive estimator, the *kernel estimator* with kernel K is defined by

$$\hat{f}(x) = \frac{1}{nh} \sum_{i=1}^{n} K\left(\frac{x - X_i}{h}\right) \tag{2.2a}$$

where h is the *window width*, also called the *smoothing parameter* or *bandwidth* by some authors. We shall consider some mathematical properties of the kernel estimator later, but first of all an intuitive discussion with some examples may be helpful.

Just as the naive estimator can be considered as a sum of 'boxes' centred at the observations, the kernel estimator is a sum of 'bumps' placed at the observations. The kernel function K determines the shape of the bumps while the window width h determines their width. An illustration is given in Fig. 2.4, where the individual bumps $n^{-1}h^{-1}K\{(x - X_i)/h\}$ are shown as well as the estimate $\hat{f}$ constructed by adding them up. It should be stressed that it is not usually appropriate to construct a density estimate from such a small sample, but that a sample of size 7 has been used here for the sake of clarity.

The effect of varying the window width is illustrated in Fig. 2.5. The limit as h tends to zero is (in a sense) a sum of Dirac delta function spikes at the observations, while as h becomes large, all detail, spurious or otherwise, is obscured.

Another illustration of the effect of varying the window width is given in Fig. 2.6. The estimates here have been constructed from a pseudo-random sample of size 200 drawn from the bimodal density given in Fig. 2.7. A normal kernel has been used to construct the estimates. Again it should be noted that if h is chosen too small then spurious fine structure becomes visible, while if h is too large then the bimodal nature of the distribution is obscured. A kernel estimate for the Old Faithful data is given in Fig. 2.8. Note that the same broad features are visible as in Fig. 2.3 but the local roughness has been eliminated.

Some elementary properties of kernel estimates follow at once from the definition. Provided the kernel K is everywhere non-negative and satisfies the condition (2.2) – in other words is a probability density function – it will follow at once from the definition that $\hat{f}$ will itself be a probability density. Furthermore, $\hat{f}$ will inherit all the continuity and differentiability properties of the kernel K, so that if, for example,

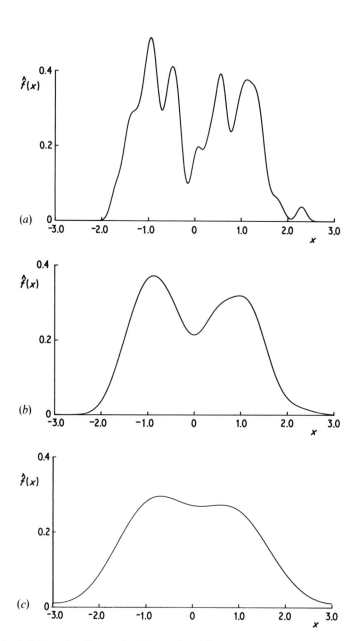

Fig. 2.6 *Kernel estimates for 200 simulated data points drawn from a bimodal density. Window widths: (a) 0.1; (b) 0.3; (c) 0.6.*

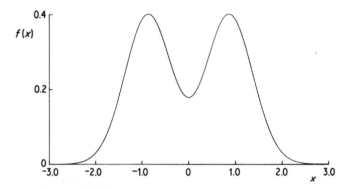

Fig. 2.7 *True bimodal density underlying data used in Fig. 2.6.*

K is the normal density function, then $\hat{f}$ will be a smooth curve having derivatives of all orders. There are arguments for sometimes using kernels which take negative as well as positive values, and these will be discussed in Section 3.6. If such a kernel is used, then the estimate may itself be negative in places. However, for most practical purposes non-negative kernels are used.

Apart from the histogram, the kernel estimator is probably the most commonly used estimator and is certainly the most studied mathematically. It does, however, suffer from a slight drawback when

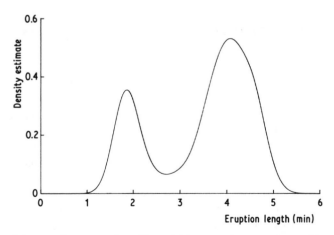

Fig. 2.8 *Kernel estimate for Old Faithful geyser data, window width 0.25.*

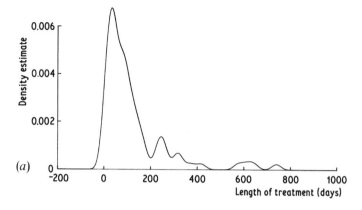

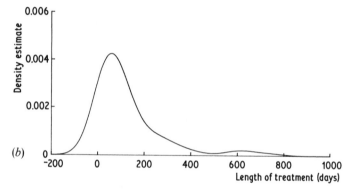

Fig. 2.9 *Kernel estimates for suicide study data. Window widths: (a) 20;*
(b) 60.

applied to data from long-tailed distributions. Because the window
width is fixed across the entire sample, there is a tendency for spurious
noise to appear in the tails of the estimates; if the estimates are
smoothed sufficiently to deal with this, then essential detail in the
main part of the distribution is masked. An example of this behaviour
is given by disregarding the fact that the suicide data are naturally
non-negative and estimating their density treating them as observations on $(-\infty, \infty)$. The estimate shown in Fig. 2.9(a) with window
width 20 is noisy in the right-hand tail, while the estimate (b) with
window width 60 still shows a slight bump in the tail and yet
exaggerates the width of the main bulge of the distribution. In order to

deal with this difficulty, various adaptive methods have been proposed, and these are discussed in the next two sections. A detailed consideration of the kernel method for univariate data will be given in Chapter 3, while Chapter 4 concentrates on the generalization to the multivariate case.

2.5 The nearest neighbour method

The nearest neighbour class of estimators represents an attempt to adapt the amount of smoothing to the 'local' density of data. The degree of smoothing is controlled by an integer k, chosen to be considerably smaller than the sample size; typically $k \approx n^{1/2}$. Define the distance $d(x, y)$ between two points on the line to be $|x - y|$ in the usual way, and for each t define

$$d_1(t) \leqslant d_2(t) \leqslant \ldots \leqslant d_n(t)$$

to be the distances, arranged in ascending order, from t to the points of the sample.

The *kth nearest neighbour density estimate* is then defined by

$$\hat{f}(t) = \frac{k}{2nd_k(t)} \qquad (2.3)$$

In order to understand this definition, suppose that the density at t is $f(t)$. Then, of a sample of size n, one would expect about $2rnf(t)$ observations to fall in the interval $[t - r, t + r]$ for each $r > 0$; see the discussion of the naive estimator in Section 2.3 above. Since, by definition, exactly k observations fall in the interval $[t - d_k(t), t + d_k(t)]$, an estimate of the density at t may be obtained by putting

$$k = 2d_k(t)n\hat{f}(t);$$

this can be rearranged to give the definition of the kth nearest neighbour estimate.

While the naive estimator is based on the number of observations falling in a box of fixed width centred at the point of interest, the nearest neighbour estimate is inversely proportional to the size of the box needed to contain a given number of observations. In the tails of the distribution, the distance $d_k(t)$ will be larger than in the main part of the distribution, and so the problem of undersmoothing in the tails should be reduced.

Like the naive estimator, to which it is related, the nearest

neighbour estimate as defined in (2.3) is not a smooth curve. The
function $d_k(t)$ can easily be seen to be continuous, but its derivative
will have a discontinuity at every point of the form $\frac{1}{2}(X_{(j)} + X_{(j+k)})$,
where $X_{(j)}$ are the order statistics of the sample. It follows at once from
these remarks and from the definition that $\hat{f}$ will be positive and
continuous everywhere, but will have discontinuous derivative at all
the same points as d_k. In contrast to the kernel estimate, the nearest
neighbour estimate will not itself be a probability density, since it
will not integrate to unity. For t less than the smallest data point, we
will have $d_k(t) = X_{(k)} - t$ and for $t > X_{(n)}$ we will have $d_k(t) =
t - X_{(n-k+1)}$. Substituting into (2.3), it follows that $\int_{-\infty}^{\infty} \hat{f}(t)\mathrm{d}t$ is
infinite and that the tails of $\hat{f}$ die away at rate t^{-1}, in other words
extremely slowly. Thus the nearest neighbour estimate is unlikely to
be appropriate if an estimate of the entire density is required.
Figure 2.10 gives a nearest neighbour density estimate for the Old
Faithful data. The heavy tails and the discontinuities in the derivative
are clear.

It is possible to generalize the nearest neighbour estimate to
provide an estimate related to the kernel estimate. As in Section 2.4,
let $K(x)$ be a kernel function integrating to one. Then the *generalized*

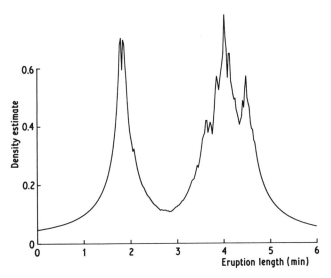

Fig. 2.10 *Nearest neighbour estimate for Old Faithful geyser data, $k = 20$.*

kth nearest neighbour estimate is defined by

$$\hat{f}(t) = \frac{1}{nd_k(t)} \sum_{i=1}^{n} K\left(\frac{t - X_i}{d_k(t)}\right). \tag{2.4}$$

It can be seen at once that $\hat{f}(t)$ is precisely the kernel estimate evaluated at t with window width $d_k(t)$. Thus the overall amount of smoothing is governed by the choice of the integer k, but the window width used at any particular point depends on the density of observations near that point.

The ordinary kth nearest neighbour estimate is the special case of (2.4) when K is the uniform kernel w of (2.1); thus (2.4) stands in the same relation to (2.3) as the kernel estimator does to the naive estimator. However, the derivative of the generalized nearest neighbour estimate will be discontinuous at all the points where the function $d_k(t)$ has discontinuous derivative. The precise integrability and tail properties will depend on the exact form of the kernel, and will not be discussed further here.

Further discussion of the nearest neighbour approach will be given in Section 5.2.

2.6 The variable kernel method

The variable kernel method is somewhat related to the nearest neighbour approach and is another method which adapts the amount of smoothing to the local density of data. The estimate is constructed similarly to the classical kernel estimate, but the scale parameter of the 'bumps' placed on the data points is allowed to vary from one data point to another.

Let K be a kernel function and k a positive integer. Define $d_{j,k}$ to be the distance from X_j to the kth nearest point in the set comprising the other $n-1$ data points. Then the *variable kernel estimate* with smoothing parameter h is defined by

$$\hat{f}(t) = \frac{1}{n} \sum_{j=1}^{n} \frac{1}{hd_{j,k}} K\left(\frac{t - X_j}{hd_{j,k}}\right). \tag{2.5}$$

The window width of the kernel placed on the point X_j is proportional to $d_{j,k}$, so that data points in regions where the data are sparse will have flatter kernels associated with them. For any fixed k, the overall degree of smoothing will depend on the parameter h. The choice of k determines how responsive the window width choice will be to very local detail.

Some comparison of the variable kernel estimate with the generalized nearest neighbour estimate (2.4) may be instructive. In (2.4) the window width used to construct the estimate at t depends on the distances from t to the data points; in (2.5) the window widths are independent of the point t at which the density is being estimated, and depend only on the distances between the data points.

In contrast with the generalized nearest neighbour estimate, the variable kernel estimate will itself be a probability density function provided the kernel K is; that is an immediate consequence of the definition. Furthermore, as with the ordinary kernel estimator, all the local smoothness properties of the kernel will be inherited by the estimate. In Fig. 2.11 the method is used to obtain an estimate for the suicide data. The noise in the tail of the curve has been eliminated, but it is interesting to note that the method exposes some structure in the main part of the distribution which is not really visible even in the undersmoothed curve in Figure 2.9.

In Section 5.3, the variable kernel method will be considered in

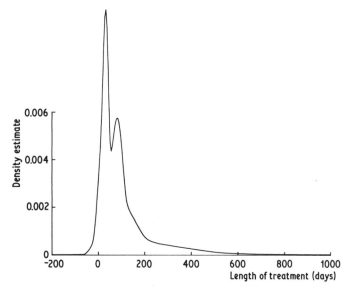

Fig. 2.11 *Variable kernel estimate for suicide study data, $k = 8$, $h = 5$.*

greater detail, and in particular an important generalization, called the adaptive kernel method, will be introduced.

2.7 Orthogonal series estimators

Orthogonal series estimators approach the density estimation problem from quite a different point of view. They are best explained by a specific example. Suppose that we are trying to estimate a density f on the unit interval $[0, 1]$. The idea of the orthogonal series method is then to estimate f by estimating the coefficients of its Fourier expansion.

Define the sequence $\phi_v(x)$ by

$$\phi_0(x) = 1$$
$$\left.\begin{array}{l} \phi_{2r-1}(x) = \sqrt{2}\cos 2\pi r x \\ \phi_{2r}(x) = \sqrt{2}\sin 2\pi r x \end{array}\right\} \quad r = 1, 2, \ldots$$

Then, by standard mathematical analysis, f can be represented as the Fourier series $\Sigma_{v=0}^{\infty} f_v \phi_v$, where, for each $v \geq 0$,

$$f_v = \int_0^1 f(x)\phi_v(x)\,\mathrm{d}x. \tag{2.6}$$

For a discussion of the sense in which f is represented by the series, see, for example, Kreider *et al.* (1966).

Suppose X is a random variable with density f. Then (2.6) can be written

$$f_v = E\phi_v(X)$$

and hence a natural, and unbiased, estimator of f_v based on a sample $X_1, \ldots, X_n$ from f is

$$\hat{f}_v = \frac{1}{n}\sum_{i=1}^{n} \phi_v(X_i).$$

Unfortunately, the sum $\Sigma_{v=0}^{\infty}\hat{f}_v\phi_v$ will not be a good estimate of f, but will 'converge' to a sum of delta functions at the observations; to see this, let

$$\omega(x) = \frac{1}{n}\sum_{i=1}^{n} \delta(x - X_i) \tag{2.7}$$

where δ is the Dirac delta function. Then, for each v,

$$\hat{f}_v = \int_0^1 \omega(x)\phi_v(x)\,dx$$

and so the $\hat{f}_v$ are exactly the Fourier coefficients of the function ω.

In order to obtain a useful estimate of the density f it is necessary to smooth ω by applying a low-pass filter to the sequence of coefficients $\hat{f}_v$. The easiest way to do this is to truncate the expansion $\Sigma \hat{f}_v \phi_v$ at some point. Choose an integer K and define the density estimate $\hat{f}$ by

$$\hat{f}(x) = \sum_{v=0}^K \hat{f}_v \phi_v(x). \tag{2.8}$$

The choice of the cutoff point K determines the amount of smoothing.

A more general approach is to taper the series by a sequence of weights λ_v, which satisfy $\lambda_v \to 0$ as $v \to \infty$, to obtain the estimate

$$\hat{f}(x) = \sum_{v=0}^{\infty} \lambda_v \hat{f}_v \phi_v(x).$$

The rate at which the weights λ_v converge to zero will determine the amount of smoothing.

Other orthogonal series estimates, no longer necessarily confined to data lying on a finite interval, can be obtained by using different orthonormal sequences of functions. Suppose $a(x)$ is a weighting function and (ψ_v) is a series satisfying, for μ and $v \geqslant 0$,

$$\int_{-\infty}^{\infty} \psi_\mu(x)\psi_v(x)a(x)\,dx = \begin{cases} 1 & \mu = v \\ 0 & \text{otherwise.} \end{cases}$$

For instance, for data rescaled to have zero mean and unit variance, $a(x)$ might be the function $e^{-x^2/2}$ and the ψ_v multiples of the Hermite polynomials; for details see Kreider et al. (1966).

The sample coefficients will then be defined by

$$\hat{f}_v = \frac{1}{n}\sum_i \psi_v(X_i)a(X_i)$$

but otherwise the estimates will be defined as above; possible estimates are

$$\hat{f}(x) = \sum_{v=0}^K \hat{f}_v \psi_v(x) \tag{2.9}$$

or

$$\hat{f}(x) = \sum_{v=0}^{\infty} \lambda_v \hat{f}_v \psi_v(x). \tag{2.10}$$

The properties of estimates obtained by the orthogonal series method depend on the details of the series being used and on the system of weights. The Fourier series estimates will integrate to unity, provided $\lambda_0 = 1$, since

$$\int_0^1 \phi_v(x)\,dx = 0 \qquad \text{for all } v > 0$$

and $\hat{f}_0$ will always be equal to one. However, except for rather special choices of the weights λ_v, $\hat{f}$ cannot be guaranteed to be non-negative. The local smoothness properties of the estimates will again depend on the particular case; estimates obtained from (2.8) will have derivatives of all orders.

2.8 Maximum penalized likelihood estimators

The methods discussed so far are all derived in an *ad hoc* way from the definition of a density. It is interesting to ask whether it is possible to apply standard statistical techniques, like maximum likelihood, to density estimation. The *likelihood* of a curve g as density underlying a set of independent identically distributed observations is given by

$$L(g|X_1,\ldots,X_n) = \prod_{i=1}^{n} g(X_i).$$

This likelihood has no finite maximum over the class of all densities. To see this, let $\hat{f}_h$ be the naive density estimate with window width $\frac{1}{2}h$; then, for each i,

$$\hat{f}_h(X_i) \geqslant \frac{1}{nh}$$

and so

$$\prod \hat{f}_h(X_i) \geqslant n^{-n}h^{-n} \to \infty \qquad \text{as } h \to 0.$$

Thus the likelihood can be made arbitrarily large by taking densities approaching the sum of delta functions ω as defined in (2.7) above, and it is not possible to use maximum likelihood directly for density estimation without placing restrictions on the class of densities over which the likelihood is to be maximized.

There are, nevertheless, possible approaches related to maximum likelihood. One method is to incorporate into the likelihood a term which describes the roughness – in some sense – of the curve under consideration. Suppose $R(g)$ is a functional which quantifies the roughness of g. One possible choice of such a functional is

$$R(g) = \int_{-\infty}^{\infty} (g'')^2. \qquad (2.11)$$

Define the *penalized log likelihood* by

$$l_\alpha(g) = \sum_{i=1}^{n} \log g(X_i) - \alpha R(g) \qquad (2.12)$$

where α is a positive smoothing parameter.

The penalized log likelihood can be seen as a way of quantifying the conflict between smoothness and goodness-of-fit to the data, since the log likelihood term $\Sigma \log g(X_i)$ measures how well g fits the data. The probability density function $\hat{f}$ is said to be a *maximum penalized likelihood density estimate* if it maximizes $l_\alpha(g)$ over the class of all curves g which satisfy $\int_{-\infty}^{\infty} g = 1$, $g(x) \geq 0$ for all x, and $R(g) < \infty$. The parameter α controls the amount of smoothing since it determines the 'rate of exchange' between smoothness and goodness-of-fit; the smaller the value of α, the rougher – in terms of $R(\hat{f})$ – will be the corresponding maximum penalized likelihood estimator. Estimates obtained by the maximum penalized likelihood method will, by definition, be probability densities. Further details of these estimates will be given in Section 5.4.

2.9 General weight function estimators

It is possible to define a general class of density estimators which includes several of the estimators discussed above. Suppose that $w(x, y)$ is a function of two arguments, which in most cases will satisfy the conditions

$$\int_{-\infty}^{\infty} w(x, y)\, \mathrm{d}y = 1 \qquad (2.13)$$

and

$$w(x, y) \geq 0 \qquad \text{for all } x \text{ and } y. \qquad (2.14)$$

We should think of w as being defined in such a way that most of the weight of the probability density $w(x, \cdot)$ falls near x. An estimate of the density underlying the data may be obtained by putting

$$\hat{f}(t) = \frac{1}{n} \sum_{i=1}^{n} w(X_i, t).$$ (2.15)

We shall refer to estimates of the form (2.15) as *general weight function estimates*. It is clear from (2.15) that the conditions (2.13) and (2.14) will be sufficient to ensure that $\hat{f}$ is a probability density function, and that the smoothness properties of $\hat{f}$ will be inherited from those of the functions $w(x, \cdot)$. This class of estimators can be thought of in two ways. Firstly, it is a unifying concept which makes it possible, for example, to obtain theoretical results applicable to a whole range of apparently distinct estimators. On the other hand, it is possible to define useful estimators which do not fall into any of the classes discussed in previous sections but which are nevertheless of the form (2.15). We shall discuss such an estimator later in this section.

To obtain the histogram as a special case of (2.15), set

$$w(x, y) = \begin{cases} \dfrac{1}{h(x)} & \text{if } x \text{ and } y \text{ fall in the same bin} \\[2ex] 0 & \text{otherwise,} \end{cases}$$

where $h(x)$ is the width of the bin containing x.

The kernel estimate can be obtained by putting

$$w(x, y) = \frac{1}{h} K\left(\frac{y - x}{h}\right).$$ (2.15a)

The orthogonal series estimate as defined in (2.8) above is given by putting

$$w(x, y) = \sum_{v=0}^{K} \phi_v(x)\phi_v(y);$$

the generalization (2.10) is obtained from

$$w(x, y) = \sum_{v=0}^{\infty} \lambda_v a(x)\psi_v(x)\psi_v(y).$$

Another example of a general weight function estimator can be obtained by considering how we would deal with data which lie naturally on the positive half-line, a topic which will be discussed at

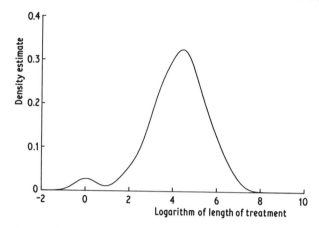

Fig. 2.12 *Kernel estimate for logarithms of suicide study data, window width 0.5.*

greater length in Section 2.10. One way of dealing with such data is to use a weight function which is, for each fixed x, a probability density which has support on the positive half-line and which has its mass concentrated near x. For example, one could choose $w(x, \cdot)$ to be a gamma density with mean x or a log-normal density with median x; in both cases, the amount of smoothing would be controlled by the

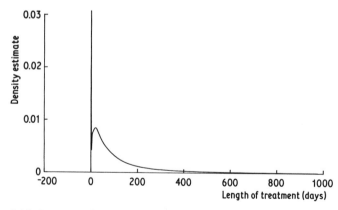

Fig. 2.13 *Log-normal weight function estimate for suicide study data, obtained by transformation of Fig. 2.12. Note that the vertical scale differs from that used in previous figures for this data set.*

choice of the shape parameter. It should be stressed that the densities $w(x, \cdot)$ will become progressively more concentrated as x approaches zero and hence the amount of smoothing applied near zero will be much less than in the right-hand tail. Using the log-normal weight function corresponds precisely to applying the kernel method, with normal kernel, to the logarithms of the data points, and then performing the appropriate inverse transformation.

An example for which this treatment is clearly appropriate is the suicide data discussed earlier. Figure 2.12 gives a kernel estimate of the density underlying the logarithms of the data values; the corresponding density estimate for the raw data is given in Fig. 2.13. The relative undersmoothing near the origin is made abundantly clear by the large spike in the estimate.

2.10 Bounded domains and directional data

It is very often the case that the natural domain of definition of a density to be estimated is not the whole real line but an interval bounded on one or both sides. For example, both the suicide data and the Old Faithful eruption lengths are measurements of positive quantities, and so it will be preferable for many purposes to obtain density estimates $\hat{f}$ for which $\hat{f}(x)$ is zero for all negative x. In the case of the Old Faithful data, the problem is really of no practical importance, since there are no observations near zero, and so the left-hand boundary can simply be ignored. The suicide data are of course quite another matter. For exploratory purposes it will probably suffice to ignore the boundary condition, but for other applications, and for presentation of the data, estimates which give any weight to the negative numbers are likely to be unacceptable.

One possible way of ensuring that $\hat{f}(x)$ is zero for negative x is simply to calculate the estimate for positive x ignoring the boundary conditions, and then to set $\hat{f}(x)$ to zero for negative x. A drawback of this approach is that if we use a method, for example the kernel method, which usually produces estimates which are probability densities, the estimates obtained will no longer integrate to unity. To make matters worse, the contribution to $\int_0^\infty \hat{f}(x)\,dx$ of points near zero will be much less than that of points well away from the boundary, and so, even if the estimate is rescaled to make it a probability density, the weight of the distribution near zero will be underestimated.

Some of the methods can be adapted to deal directly with data on the half-line. For example, we could use an orthogonal series estimate of the form (2.9) or (2.10) with functions ψ_ν which were orthonormal with respect to a weighting function a which is zero for $x < 0$. The maximum penalized likelihood method can be adapted simply by constraining $g(x)$ to be zero for negative x, and using a roughness penalty functional which only depends on the behaviour of g on $(0, \infty)$.

Another possible approach is to transform the data, for example by taking logarithms as in the example given in Section 2.9 above. If the density estimated from the logarithms of the data is $\hat{g}$, then standard arguments lead to

$$\hat{f}(x) = \frac{1}{x} \, \hat{g}(\log x) \qquad \text{for } x > 0.$$

It is of course the presence of the multiplier $1/x$ that gives rise to the spike in Fig. 2.13; notwithstanding difficulties of this kind, Copas and Fryer (1980) did find estimates based on logarithmic transforms to be very useful with some other data sets.

It is possible to use other adaptations of methods originally designed for the whole real line. Suppose we augment the data by adding the reflections of all the points in the boundary, to give the set $\{X_1, -X_1, X_2, -X_2, \ldots\}$. If a kernel estimate f^* is constructed from this data set of size $2n$, then an estimate based on the original data can be given by putting

$$\hat{f}(x) = \begin{cases} 2f^*(x) & \text{for } x \geq 0 \\ 0 & \text{for } x < 0. \end{cases}$$

This estimate corresponds to a general weight function estimator with, for x and $y > 0$,

$$w(x, y) = \frac{1}{h} K\left(\frac{y - x}{h}\right) + \frac{1}{h} K\left(\frac{y + x}{h}\right).$$

Provided the kernel is symmetric and differentiable, some easy manipulation shows that the estimate will always have zero derivative at the boundary. If the kernel is a symmetric probability density, the estimate will be a probability density. It is clear that it is not usually necessary to reflect the whole data set, since if X_i/h is sufficiently large, the reflected point $-X_i/h$ will not be felt in the calculation of $f^*(x)$ for $x \geq 0$, and hence we need only reflect points near 0. For example, if K is the normal kernel there is no practical need to reflect points $X_i > 4h$.

This reflection technique can be used in conjunction with any method for density estimation on the whole line. With most methods estimates which satisfy $\hat{f}'(0+) = 0$ will be obtained.

Another, related, technique forces $\hat{f}(0+) = 0$ rather than $\hat{f}'(0+) = 0$. Reflect the data as before, but give the reflected points weight -1 in the calculation of the estimate; thus the estimate is, for $x \geqslant 0$,

$$\hat{f}(x) = \frac{1}{nh} \sum_{i=1}^{n} \left[K\left(\frac{x-X_i}{h}\right) - K\left(\frac{x+X_i}{h}\right) \right]. \qquad (2.16)$$

We shall call this technique *negative reflection*. Estimates constructed from (2.16) will no longer integrate to unity, and indeed the total contribution to $\int_0^\infty \hat{f}(x)\,dx$ from points near the boundary will be small. Whether estimates of this form are useful depends on the context.

All the remarks of this section can be extended to the case where the required support of the estimator is a finite interval $[a,b]$. Transformation methods can be based on transformations of the form

$$Y_i = H^{-1}\left(\frac{X_i - a}{b - a}\right)$$

where H is any cumulative probability distribution function strictly increasing on $(-\infty, \infty)$. Generally, the estimates obtained by transformation back to the original scale will be less smoothed for points near the boundaries. The reflection methods are easily generalized. It is necessary to reflect in both boundaries and it is of course possible to use ordinary reflection in one boundary and negative reflection in the other, if the corresponding boundary conditions are required.

Another way of dealing with data on a finite interval $[a,b]$ is to impose periodic or 'wrap around' boundary conditions. Of course this approach is particularly useful if the data are actually directions or angles; the turtle data considered in Section 1.2 were of this kind. For simplicity, suppose that the interval on which the data naturally lie is $[0, 1]$, which can be regarded as a circle of circumference 1; more general intervals are dealt with analogously. If we want to use a method like the kernel method, a possible approach is to wrap the kernel round the circle. Computationally it may be simpler to augment the data set by replicating shifted copies of it on the intervals

$[-1, 0]$ and $[1, 2]$, to obtain the set

$$\{X_1 - 1, X_2 - 1, \ldots, X_n - 1, X_1, X_2, \ldots, X_n, X_1 + 1,$$
$$X_2 + 1, \ldots, X_n + 1\}. \tag{2.17}$$

In principle we should continue to replicate on intervals further away from $[0, 1]$, but that is rarely necessary in practice. Applying the kernel method or one of its variants to the augmented data set will give an estimate on $[0, 1]$ which has the required boundary property; of course the factor $1/n$ should be retained in the definition of the estimate even though the augmented data set has more than n members.

The orthogonal series estimates based on Fourier series will automatically impose periodic boundary conditions, because of the periodicity of the functions ϕ_v of section 2.7.

2.11 Discussion and bibliography

A brief survey of the kind conducted in this chapter of course asks far more questions than it answers, and some of these questions will be the subject of discussion in subsequent chapters. The overriding problems are the choice of what method to use in any given practical context and, given that a particular method is being used, how to choose the various parameters needed by the method. The remarks already made about the mathematical properties of the estimates obtained by various procedures will of course be important in making these decisions. To obtain a fuller understanding of the importance and consequences of the various choices it is essential to investigate the statistical properties of the various methods and also to consider the difficulties involved in computing the estimates.

This chapter has by no means considered all the methods available for density estimation. Generalizations and other approaches are considered in later chapters of this book, and in the other books and surveys mentioned in Section 1.3.

The naive estimator was introduced by Fix and Hodges (1951) in an unpublished report; the first published paper to deal explicitly with probability density estimation was by Rosenblatt (1956), who discussed both the naive estimator and the more general kernel estimator. Whittle (1958) formulated the general weight function class of estimators, while the orthogonal series estimator was introduced by Čencov (1962). The nearest neighbour estimate was first considered by Loftsgaarden and Quesenberry (1965), while the variable kernel method is due to Breiman, Meisel and Purcell (1977), though Wertz (1978, p. 59) refers to presumably independent but related work by Victor. The maximum penalized likelihood approach was first applied to density estimation by Good and Gaskins (1971). The reflection and replication techniques of Section 2.10 were introduced and illustrated by Boneva, Kendall and Stefanov (1971), while the transformation technique is discussed by Copas and Fryer (1980).

CHAPTER 3

The kernel method for univariate data

3.1 Introduction

In this chapter the elementary statistical properties of the kernel estimator in the univariate case will be discussed in more detail. Our concentration on the kernel method is not intended to imply that the method is the best to use in all circumstances, but there are several reasons for considering the kernel method first of all. The method is of wide applicability, particularly in the univariate case, and it is certainly worth understanding its behaviour before going on to consider other methods. It is probably the method whose properties are best understood, and discussion of these properties raises issues which relate to other methods of density estimation.

The development will include both theoretical and practical aspects of the method, but, generally, only that theory of reasonably immediate practical relevance will be included.

3.1.1 *Notation and conventions*

Throughout this chapter it will be assumed that we have a sample $X_1, \ldots, X_n$ of independent, identically distributed observations from a continuous univariate distribution with probability density function f, which we are trying to estimate. There are many practical problems where these assumptions are not necessarily justifiable, but nevertheless they provide a standard framework in which to discuss the properties of density estimation methods.

Throughout this chapter, except where otherwise stated, $\hat{f}$ will be the kernel estimator with kernel K and window width h, as defined and discussed in Section 2.4.

The basic methodology of the theoretical treatment is to discuss the closeness of the estimator $\hat{f}$ to the true density f in various senses. The estimate $\hat{f}$ of course depends on the data as well as on the kernel and the window width; this dependence will not generally be expressed explicitly. For each x, $\hat{f}(x)$ can be thought of as a random variable, because of its dependence on the observations $X_1, \ldots, X_n$; any use of probability, expectation and variance involving $\hat{f}$ is with respect to its sampling distribution as a statistic based on these random observations.

Except where otherwise stated Σ will refer to a sum for $i = 1$ to n and $\int$ to an integral over the range $(-\infty, \infty)$.

3.1.2 Measures of discrepancy : mean square error and mean integrated square error

Various measures have been studied of the discrepancy of the density estimator $\hat{f}$ from the true density f. When considering estimation at a single point, a natural measure is the *mean square error* (abbreviated MSE), defined by

$$\mathrm{MSE}_x(\hat{f}) = E\{\hat{f}(x) - f(x)\}^2. \qquad (3.1)$$

By standard elementary properties of mean and variance,

$$\mathrm{MSE}_x(\hat{f}) = \{E\hat{f}(x) - f(x)\}^2 + \mathrm{var}\,\hat{f}(x), \qquad (3.2)$$

the sum of the squared bias and the variance at x. We shall see that, as in many branches of statistics, there is a trade-off between the bias and variance terms in (3.2); the bias can be reduced at the expense of increasing the variance, and vice versa, by adjusting the amount of smoothing.

The first (Rosenblatt, 1956) and the most widely used way of placing a measure on the *global* accuracy of $\hat{f}$ as an estimator of f is the *mean integrated square error* (abbreviated MISE) defined by

$$\mathrm{MISE}(\hat{f}) = E \int \{\hat{f}(x) - f(x)\}^2 \, \mathrm{d}x. \qquad (3.3)$$

Though there are other global measures of discrepancy which may be more appropriate to one's intuitive ideas about what constitutes a globally good estimate, the MISE is by far the most tractable global measure, and so is well worth studying first of all. Other measures of discrepancy will be discussed in Section 3.7.

It is useful to note that, since the integrand is non-negative, the order of integration and expectation in (3.3) can be reversed to give the alternative forms

$$\text{MISE}(\hat{f}) = \int E\{\hat{f}(x) - f(x)\}^2 \, dx$$

$$= \int \text{MSE}_x(\hat{f}) \, dx \qquad (3.4)$$

$$= \int \{E\hat{f}(x) - f(x)\}^2 \, dx + \int \text{var}\,\hat{f}(x) \, dx, \qquad (3.5)$$

which gives the MISE as the sum of the *integrated* square bias and the *integrated* variance.

3.2 Elementary finite sample properties

Suppose $\hat{f}$ is the general weight function estimate defined in (2.15) above. By elementary manipulations (Whittle, 1958) it follows that, for each t,

$$E\hat{f}(t) = \frac{1}{n} \sum E w(X_i, t) = \int w(x, t) f(x) \, dx \qquad (3.6)$$

and since the X_i are independent,

$$\text{var}\,\hat{f}(t) = \frac{1}{n} \text{var}\, w(X_i, t)$$

$$= \frac{1}{n} \left[\int w(x, t)^2 f(x) \, dx - \left\{ \int w(x, t) f(x) \, dx \right\}^2 \right]. \qquad (3.7)$$

Expressions for the MSE and MISE can be obtained by substituting these formulae into (3.5). A very interesting property of (3.6) is that for given f, the bias $E\hat{f}(t) - f(t)$ does not depend on the sample size directly, but depends only on the weight function. This is important conceptually because it shows that taking larger and larger samples will not, alone, reduce the bias; it will be necessary to adjust the weight function to obtain asymptotically unbiased estimates.

3.2.1 *Application to kernel estimates*

Substituting formula (2.15a) gives the kernel estimate as a general weight function estimate. Therefore, using $\hat{f}$ to denote the kernel estimate, the following can be obtained from (3.6) and (3.7):

$$E \hat{f}(x) = \int \frac{1}{h} K\left(\frac{x-y}{h}\right) f(y) \, dy \qquad (3.8)$$

$$n \operatorname{var} \hat{f}(x) = \int \frac{1}{h^2} K\left(\frac{x-y}{h}\right)^2 f(y) \, dy$$
$$- \left\{ \frac{1}{h} \int K\left(\frac{x-y}{h}\right) f(y) \, dy \right\}^2 . \qquad (3.9)$$

Again it is straightforward, in principle, to substitute these expressions into (3.2) and (3.5) to obtain exact expressions for the MSE and MISE, but except in very special cases the calculations become intractable and the expressions obtained have little intuitive meaning. It is more instructive to obtain approximations to (3.8) and (3.9) under suitable conditions, and this will be done in Section 3.3.

It can be seen from (3.8) that the expected value of $\hat{f}$ is precisely a smoothed version of the true density, obtained by convolving f with the kernel scaled by the window width. It is characteristic of almost all density estimation methods that the estimate is of the form

smoothed version of true density + random error   (3.10)

where the 'smoothed version of the true density' depends deterministically on the precise choice of parameters in the method, but not directly on the sample size.

One very special case where (3.8) and (3.9) are reasonably tractable is the case where the kernel is the standard normal density and the true density is also normal, with mean μ and variance σ^2. For full details of the calculations see Fryer (1976) or Deheuvels (1977). It is easily seen that $E\hat{f}$ is the $N(\mu, \sigma^2 + h^2)$ density. Straightforward but long-winded calculations then yield an expression for $\mathrm{MSE}_x(\hat{f})$ as a weighted sum of normal density functions (equation 4 of p. 373 of Fryer, 1976); this expression can then be integrated to obtain

$$(2\sqrt{\pi})\mathrm{MISE} = n^{-1}\{h^{-1} - (\sigma^2 + h^2)^{-1/2}\}$$
$$+ \sigma^{-1} + (\sigma^2 + h^2)^{-1/2} - 2\sqrt{2}(2\sigma^2 + h^2)^{-1/2}. \qquad (3.11)$$

The basic tricks used in the calculations are the facts that the convolution of normal densities is normal, and that the square of a normal density is a multiple of another normal density. Deheuvels (1977) also considers some other special cases.

The expression (3.11) can be minimized over h to find the optimal window width. It turns out that the results obtained are, even for very small sample sizes, almost identical with the more tractable asymptotic formula (3.21) given in Section 3.3 below. The tables on p. 39 of Deheuvels (1977) show that the approximations are good even for samples of size 10.

3.3 Approximate properties

Very many of the theoretical papers written on density estimation deal with asymptotic properties of the various methods and, as in many other branches of mathematical statistics, there has been something of an emphasis on limit theory for its own sake. Nevertheless, asymptotic results are not without important practical uses, one of which is qualitative, in providing intuition about the way that techniques behave without having to grasp formulae like (3.11) above.

In the remainder of this section we shall derive approximate expressions for the bias and variance, and use these to investigate how the mean square error and mean integrated square error will behave. Later on, in Section 3.7, we shall refer to rigorous asymptotic results which justify these approximations, under suitable conditions. For simplicity, we shall assume throughout this discussion that:

the kernel K is a symmetric function satisfying

$$\int K(t)\,dt = 1, \qquad \int tK(t)\,dt = 0, \quad \text{and} \quad \int t^2 K(t)\,dt = k_2 \neq 0 \quad (3.12)$$

and that

the unknown density f has continuous derivatives of all orders required. $\hspace{1cm}$ (3.13)

Usually, the kernel K will be a symmetric probability density function, for example the normal density, and the constant k_2 will then be the variance of the distribution with this density. It should be stressed that, unlike the density f, the kernel K is under the user's control, and therefore it is only necessary for practical purposes to consider results which hold for the particular kernel being used.

3.3.1 *The bias and variance*

As has already been pointed out, the bias in the estimation of $f(x)$ does not depend directly on the sample size, but does depend on the window width h. Of course, if h is chosen as a function of n, then the bias will depend indirectly on n through its dependence on h. We shall write

$$\text{bias}_h(x) = E\hat{f}(x) - f(x)$$
$$= \int h^{-1}K\{(x-y)/h\}f(y)\,\mathrm{d}y - f(x). \tag{3.14}$$

The bias also depends on the kernel K, but this dependence will not be expressed explicitly. We shall now use (3.14) to obtain an approximate expression for the bias.

Make the change of variable $y = x - ht$ and use the assumption that K integrates to unity, to write

$$\text{bias}_h(x) = \int K(t)f(x-ht)\,\mathrm{d}t - f(x)$$

$$= \int K(t)\{f(x-ht) - f(x)\}\,\mathrm{d}t.$$

A Taylor series expansion gives

$$f(x-ht) = f(x) - htf'(x) + \tfrac{1}{2}h^2t^2f''(x) + \cdots$$

so that, by the assumptions made about K,

$$\text{bias}_h(x) = -hf'(x)\int tK(t)\,\mathrm{d}t + \tfrac{1}{2}h^2f''(x)\int t^2K(t)\,\mathrm{d}t + \cdots \tag{3.15}$$

$$= \tfrac{1}{2}h^2f''(x)k_2 + \text{higher-order terms in } h. \tag{3.16}$$

The integrated square bias, required in formula (3.5) for the mean integrated square error, is then given by

$$\int \text{bias}_h(x)^2\mathrm{d}x \approx \tfrac{1}{4}h^4k_2^2\int f''(x)^2\,\mathrm{d}x. \tag{3.17}$$

We now turn to the variance. From (3.9) and (3.8) we have

$$\text{var } \hat{f}(x) = n^{-1}\int h^{-2}K\{(x-y)h^{-1}\}^2f(y)\,\mathrm{d}y$$

$$- n^{-1}\{f(x) + \text{bias}_h(x)\}^2$$

$$\approx n^{-1}h^{-1}\int f(x-ht)K(t)^2\,\mathrm{d}t - n^{-1}\{f(x) + O(h^2)\}^2,$$

using the substitution $y = x - ht$ in the integral, and the approximation (3.16) for the bias. Assume that h is small and n is large, and expand $f(x - ht)$ as a Taylor series to obtain

$$\text{var } \hat{f}(x) \approx n^{-1}h^{-1} \int \{ f(x) - ht f'(x) + \cdots \} K(t)^2 \, dt + O(n^{-1})$$

$$= n^{-1}h^{-1} f(x) \int K(t)^2 \, dt + O(n^{-1})$$

$$\approx n^{-1}h^{-1} f(x) \int K(t)^2 \, dt \tag{3.18}$$

Since f is a probability density function, integrating (3.18) over x gives the simple approximation

$$\int \text{var } \hat{f}(x) \, dx \approx n^{-1}h^{-1} \int K(t)^2 \, dt. \tag{3.19}$$

Suppose, now, that we want to choose h to make the mean integrated square error as small as possible. Comparing approximations (3.17) and (3.19) for the two components of the mean integrated square error demonstrates one of the fundamental problems of density estimation. If, in an attempt to eliminate the bias, a very small value of h is used, then the integrated variance will become large. On the other hand, choosing a large value of h will reduce the random variation as quantified by the variance, at the expense of introducing systematic error, or bias, into the estimation. This discussion provides a mathematical explanation of the behaviour illustrated in Figures 2.6 and 2.7. It should be stressed that, whatever method of density estimation is being used, the choice of smoothing parameter implies a trade-off between random and systematic error.

3.3.2 The ideal window width and kernel

The ideal value of h, from the point of view of minimizing the approximate mean integrated square error

$$\tfrac{1}{4}h^4k_2{}^2 \int f''(x)^2 \, dx + n^{-1}h^{-1} \int K(t)^2 \, dt, \tag{3.20}$$

can be shown by simple calculus (Parzen, 1962, Lemma 4A) to be equal to h_{opt}, where

$$h_{\text{opt}} = k_2{}^{-2/5} \left\{ \int K(t)^2 \, dt \right\}^{1/5} \left\{ \int f''(x)^2 \, dx \right\}^{-1/5} n^{-1/5}. \tag{3.21}$$

The formula (3.21) for the optimal window width is somewhat disappointing since it shows that h_{opt} itself depends on the unknown density being estimated. Nevertheless, some useful conclusions can be drawn. Firstly, the ideal window width will converge to zero as the sample size increases, but at a very slow rate. Secondly, since the term $\int f''^2$ measures, in a sense, the rapidity of fluctuations in the density f, it can be seen from (3.21) that, as one would expect, smaller values of h will be appropriate for more rapidly fluctuating densities. A natural approach leading on from (3.21) is to choose h with reference to some standard family of densities, such as the normal densities. This idea will be explored in Section 3.4.2.

Substituting the value of h_{opt} from (3.21) back into the approximate formula (3.20) for the mean integrated square error shows that, if h is chosen optimally, then the approximate value of the mean integrated square error will be

$$\tfrac{5}{4} C(K) \left\{ \int f''(x)^2 \, \mathrm{d}x \right\}^{1/5} n^{-4/5} \tag{3.22}$$

where the constant $C(K)$ is given by

$$C(K) = k_2^{2/5} \left\{ \int K(t)^2 \, \mathrm{d}t \right\}^{4/5}. \tag{3.23}$$

Studying formula (3.22) shows that, all other things being equal, we should choose a kernel K with a small value of $C(K)$, since this will make it theoretically possible to obtain a small value of the mean integrated square error if we can choose the smoothing parameter correctly.

Focus attention, for the moment, on kernels which are themselves probability density functions. These are, in any case, the only kernels which will ensure that the estimate $\hat{f}$ is everywhere non-negative, but in some circumstances it may be worth relaxing this requirement (see Section 3.6 below). If the value of k_2 is not equal to one, replace the kernel by the rescaled version $k_2^{1/2} K(k_2^{1/2} t)$. This will not affect the value of $C(K)$.

The problem of minimizing $C(K)$ then reduces to that of minimizing $\int K(t)^2 \, \mathrm{d}t$ subject to the constraints that $\int K(t) \, \mathrm{d}t$ and $\int t^2 K(t) \, \mathrm{d}t$ are both equal to one. In a different context, Hodges and Lehmann (1956) showed that this problem is solved by setting $K(t)$ to be the function

$$K_e(t) = \begin{cases} \dfrac{3}{4\sqrt{5}}\left(1 - \dfrac{1}{5}t^2\right) & -\sqrt{5} \leqslant t \leqslant \sqrt{5} \\ \\ 0 & \text{otherwise.} \end{cases} \qquad (3.24)$$

The notation $K_e(t)$ is used because this kernel was first suggested in the density estimation context by Epanechnikov (1969), and so it is often called the *Epanechnikov kernel*. A graph of $K_e(t)$ is given in Fig. 3.1.

We can now consider the efficiency of any symmetric kernel K by comparing it with the Epanechnikov kernel. Define the *efficiency* of K to be

$$\text{eff}(K) = \{C(K_e)/C(K)\}^{5/4} \qquad (3.25)$$

$$= \frac{3}{5\sqrt{5}}\left\{\int t^2 K(t)\,dt\right\}^{-1/2}\left\{\int K(t)^2\,dt\right\}^{-1}. \qquad (3.26)$$

The reason for the power 5/4 in (3.25) is that for large n, the mean integrated square error will be the same whether we use n observations and the kernel K or whether we use $n\,\text{eff}(K)$ observations and the kernel K_e (see Section 17.29 of Kendall and Stuart, 1973, volume 2). Some kernels and their efficiencies are given in Table 3.1. It is quite remarkable that the efficiencies obtained are so close to one; even the rectangular kernel used when constructing the naive estimator has an efficiency of nearly 0.93. Particularly bearing in mind the asymptotic nature of the arguments leading to the definition of

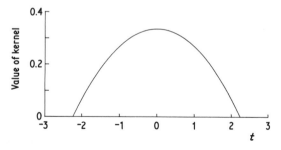

Fig. 3.1 *The Epanechnikov kernel.*

Table 3.1 *Some kernels and their efficiencies*

Kernel	$K(t)$	Efficiency (exact and to 4 d.p.)
Epanechnikov	$\frac{3}{4}(1 - \frac{1}{5}t^2)/\sqrt{5}$ for $|t| < \sqrt{5}$, 0 otherwise	1
Biweight	$\frac{15}{16}(1 - t^2)^2$ for $|t| < 1$ 0 otherwise	$\left(\dfrac{3087}{3125}\right)^{1/2} \approx 0.9939$
Triangular	$1 - |t|$ for $|t| < 1$, 0 otherwise	$\left(\dfrac{243}{250}\right)^{1/2} \approx 0.9859$
Gaussian	$\dfrac{1}{\sqrt{2\pi}} e^{-(1/2)t^2}$	$\left(\dfrac{36\pi}{125}\right)^{1/2} \approx 0.9512$
Rectangular	$\frac{1}{2}$ for $|t| < 1$, 0 otherwise	$\left(\dfrac{108}{125}\right)^{1/2} \approx 0.9295$

efficiency, the message of Table 3.1 is really that there is very little to choose between the various kernels on the basis of mean integrated square error. It is perfectly legitimate, and indeed desirable, to base the choice of kernel on other considerations, for example the degree of differentiability required or the computational effort involved (see the discussion in Section 3.5 below).

3.4 Choosing the smoothing parameter

The problem of choosing how much to smooth is of crucial importance in density estimation. Before discussing various methods in detail, it is worth pausing to make some remarks of a general nature. It should never be forgotten that the appropriate choice of smoothing parameter will always be influenced by the purpose for which the density estimate is to be used. If the purpose of density estimation is to explore the data in order to suggest possible models and hypotheses, then it will probably be quite sufficient, and indeed desirable, to choose the smoothing parameter subjectively as in Section 3.4.1 below. When using density estimation for presenting conclusions, there is a case for undersmoothing somewhat; the reader can do further smoothing 'by eye', but cannot easily unsmooth.

However, many applications require an automatic choice of

smoothing parameter. The inexperienced user will doubtless feel happier if the method is fully automatic, and an automatic choice can in any case be used as a starting point for subsequent subjective adjustment. Scientists reporting or comparing their results will want to make reference to a standardized method. If density estimation is to be used routinely on a large number of data sets or as part of a larger procedure, then an automatic method is essential.

In this discussion I have deliberately used the word *automatic* rather than *objective* for methods that do not require explicit specification of control parameters. Behind the process of automating statistical procedures completely always lies the danger of encouraging the user not to give enough consideration to prior assumptions, whether or not from a Bayesian point of view.

In the following sections, various methods for choosing the smoothing parameter are discussed. There is as yet no universally accepted approach to this problem.

3.4.1 *Subjective choice*

A natural method for choosing the smoothing parameter is to plot out several curves and choose the estimate that is most in accordance with one's prior ideas about the density. For many applications this approach will be perfectly satisfactory. Indeed, the process of examining several plots of the data, all smoothed by different amounts, may well give more insight into the data than merely considering a single automatically produced curve.

Consider, as an example, the estimate given in Fig. 3.2. The data underlying these estimates are the amounts of winter snowfall (in inches) at Buffalo, New York, for each of the 63 winters from 1910/11 to 1972/73. These data have been considered by several authors; see, for example, Parzen (1979). It can be seen from Fig. 3.2 that varying the smoothing parameter yields essentially two possible explanations of the data, either a roughly normal distribution or a trimodal curve suggesting a mixture of three populations approximately in the ratio 1:3:1. For many purposes, particularly for model and hypothesis generation, it is by no means unhelpful for the statistician to supply the scientist with a range of possible presentations of the data. A choice between the two alternative models suggested by Fig. 3.2 is a very useful step forward from the enormous number of possible explanations that could conceivably be considered.

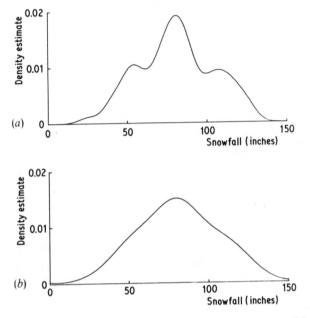

Fig. 3.2 *Kernel estimates for annual snowfall data collected at Buffalo, New York. Window widths (a) 6; (b) 12.*

3.4.2 *Reference to a standard distribution*

A very easy and natural approach is to use a standard family of distributions to assign a value to the term $\int f''(x)^2 \, dx$ in the expression (3.21) for the ideal window width. For example, the normal distribution with variance σ^2 has, setting ϕ to be the standard normal density,

$$\int f''(x)^2 \, dx = \sigma^{-5} \int \phi''(x)^2 \, dx$$
$$= \tfrac{3}{8}\pi^{-1/2}\sigma^{-5} \approx 0.212\,\sigma^{-5}. \qquad (3.27)$$

If a Gaussian kernel is being used, then the window width obtained from (3.21) would be, substituting the value (3.27),

$$h_{\text{opt}} = (4\pi)^{-1/10}(\tfrac{3}{8}\pi^{-1/2})^{-1/5}\sigma n^{-1/5}$$
$$= \left(\frac{4}{3}\right)^{1/5}\sigma n^{-1/5} = 1.06\sigma n^{-1/5}. \qquad (3.28)$$

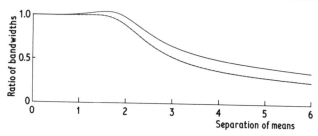

Fig. 3.3 *Ratio of asymptotically optimal window width (3.21) to window width given by rule-of-thumb. Solid curve: rule-of-thumb (3.28), based on the standard deviation; dotted curve: rule-of-thumb (3.29) based on the interquartile range. Ratios calculated for true density an equal mixture of standard normals with means separated by the given amount.*

A quick way of choosing the smoothing parameter, therefore, would be to estimate σ from the data and then to substitute the estimate into (3.28). Either the usual sample standard deviation or a more robust estimator of σ could be used.

While (3.28) will work well if the population really is normally distributed, it may oversmooth somewhat if the population is multimodal, as a result of the value of $(\int f''^2)^{1/5}$ being larger relative to the standard deviation. This effect is illustrated in Fig. 3.3, which shows the ratio of the optimum bandwidth given by (3.21) to the value obtained using (3.28) if the true f is an equal mixture of two unit normal distributions with means separated by varying amounts. The figure shows that in the range of separations (0, 2) the formula (3.28) will do extremely well; of course, the mixture density is itself unimodal throughout this range. However, as the mixture becomes more strongly bimodal the formula (3.28) will oversmooth more and more, relative to the optimal choice of smoothing parameter.

In order to investigate the sensitivity of the optimal window width to skewness and kurtosis in unimodal distributions, curves corresponding to Fig. 3.3 were calculated for the log-normal and t family of distributions. These are shown as the solid curves in Figures 3.4 and 3.5. It can be seen that, for heavily skewed data, using (3.28) will again oversmooth, but that the formula is remarkably insensitive to kurtosis within the t family of distributions.

Better results can be obtained using a robust measure of spread. Formula (3.28) written in terms of the interquartile range R of the

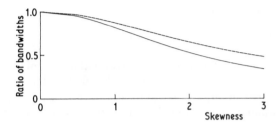

Fig. 3.4 *Ratio of asymptotically optimal window width to window widths given by rule-of-thumb, as in Fig. 3.3; for log-normal distribution with given skewness.*

underlying normal distribution becomes

$$h_{\text{opt}} = 0.79Rn^{-1/5}. \tag{3.29}$$

Using (3.29) for the long-tailed and skew distributions gives the much improved dotted curves in Figs 3.4 and 3.5. Unfortunately, using (3.29) for the bimodal distributions makes matters worse, because it oversmooths even further. The best of both worlds can be obtained using the adaptive estimate of spread

$$A = \min(\text{standard deviation, interquartile range}/1.34) \tag{3.30}$$

instead of σ in the formula (3.28). This will cope well with the unimodal densities and will not do too badly if the density is moderately bimodal. Another modification, which will improve matters further, is to reduce the factor 1.06 in (3.28); for instance, the

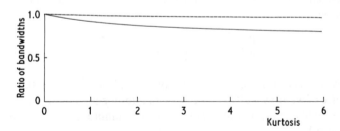

Fig. 3.5 *Ratio of asymptotically optimal window width to window widths given by rule-of-thumb, as in Fig. 3.3; for t-distribution with given kurtosis coefficient.*

choice, for a Gaussian kernel,

$$h = 0.9An^{-1/5} \tag{3.31}$$

will yield a mean integrated square error within 10% of the optimum for all the t-distributions considered, for the log-normal with skewness up to about 1.8, and for the normal mixture with separation up to 3 standard deviations; beyond these points some simulations performed by the author indicate that, for samples of size 100, the skewness or bimodality will usually be clear using the value given by (3.31), even if the density is slightly oversmoothed. In summary, the choice (3.31) for the smoothing parameter will do very well for a wide range of densities and is trivial to evaluate. For many purposes it will certainly be an adequate choice of window width, and for others it will be a good starting point for subsequent fine tuning.

Scott (1979) has provided a discussion, parallel to this section, for the problem of choosing the optimum bin width in a histogram. He makes the further suggestion that a graph like Fig. 3.4 could be used to give a correction factor for skewness in the sample.

3.4.3 *Least-squares cross-validation*

Least-squares cross-validation is a completely automatic method for choosing the smoothing parameter. It has only been formulated in recent years but is based on an extremely simple idea. The method was suggested by Rudemo (1982) and Bowman (1984). See also Bowman, Hall and Titterington (1984), Hall (1983) and Stone (1984) for further discussion.

Given any estimator $\hat{f}$ of a density f, the integrated square error can be written

$$\int (\hat{f} - f)^2 = \int \hat{f}^2 - 2 \int \hat{f}f + \int f^2. \tag{3.32}$$

Now the last term of (3.32) does not depend on $\hat{f}$, and so the ideal choice of window width (in the sense of minimizing integrated square error) will correspond to the choice which minimizes the quantity R defined by

$$R(\hat{f}) = \int \hat{f}^2 - 2 \int \hat{f}f. \tag{3.33}$$

The basic principle of least-squares cross-validation is to construct an estimate of $R(\hat{f})$ from the data themselves and then to minimize this estimate over h to give the choice of window width. The term $\int \hat{f}^2$ can be found from the estimate $\hat{f}$. Define $\hat{f}_{-i}$ to be the density estimate constructed from all the data points *except* X_i, that is to say,

$$\hat{f}_{-i}(x) = (n-1)^{-1} h^{-1} \sum_{j \neq i} K\{h^{-1}(x - X_j)\}. \qquad (3.34)$$

Now define

$$M_0(h) = \int \hat{f}^2 - 2n^{-1} \sum_i \hat{f}_{-i}(X_i). \qquad (3.35)$$

The score M_0 depends only on the data (although it is not in a very suitable form for easy calculation). The idea of least-squares cross-validation is to minimize the score M_0 over h. We shall discuss below why this procedure can be expected to give good results and also obtain a computationally simpler approximation to M_0.

In order to understand why minimizing M_0 is a sensible way to proceed, we consider the expected value of $M_0(h)$. The summation term in (3.34) has expectation

$$E n^{-1} \sum_i \hat{f}_{-i}(X_i) = E \hat{f}_{-n}(X_n)$$

$$= E \int \hat{f}_{-n}(x) f(x) \, \mathrm{d}x = E \int \hat{f}(x) f(x) \, \mathrm{d}x \qquad (3.36)$$

since $E(\hat{f})$ depends only on the kernel and the window width, and not on the sample size. Substituting (3.36) back into the definition of $M_0(h)$ shows that $EM_0(h) = ER(\hat{f})$. It follows from (3.32) that $M_0(h) + \int f^2$ is, for all h, an unbiased estimator of the mean integrated square error; since the term $\int f^2$ is the same for all h, minimizing $EM_0(h)$ corresponds precisely to minimizing the mean integrated square error. Assuming that the minimizer of M_0 is close to the minimizer of $E(M_0)$ indicates why we might hope that minimizing M_0 gives a good choice of smoothing parameter.

To express the score M_0 in a form which is more suitable for computation, first define $K^{(2)}$ to be the convolution of the kernel with itself. If, for example, K is the standard Gaussian kernel, then $K^{(2)}$ will be the Gaussian density with variance 2. Now, assuming K is

DENSITY ESTIMATION

symmetric, we have, substituting $u = h^{-1}x$,

$$
\begin{aligned}
\int \hat{f}(x)^2 dx &= \int \sum_i n^{-1} h^{-1} K\{h^{-1}(x - X_i)\} \\
&\quad \times \sum_j n^{-1} h^{-1} K\{h^{-1}(x - X_j)\} dx \\
&= n^{-2} h^{-1} \sum_i \sum_j \int K(h^{-1}X_i - u) K(u - h^{-1}X_j) du \\
&= n^{-2} h^{-1} \sum_i \sum_j K^{(2)}\{h^{-1}(X_i - X_j)\}.
\end{aligned}
\tag{3.37}
$$

Also

$$
\begin{aligned}
n^{-1} \sum f_{-i}(X_i) &= n^{-1} \sum_i (n-1)^{-1} \sum_{j \neq i} h^{-1} K\{h^{-1}(X_i - X_j)\} \\
&= n^{-1}(n-1)^{-1} \sum_i \sum_j h^{-1} K\{h^{-1}(X_i - X_j)\} - (n-1)^{-1} h^{-1} K(0)
\end{aligned}
\tag{3.38}
$$

To find $M_0(h)$, the expressions (3.37) and (3.38) can be substituted into the definition (3.35). A very closely related score function $M_1(h)$, still easier to calculate, is given by changing the factors $(n-1)^{-1}$ in (3.38) to the simpler n^{-1}, and then substituting into (3.35) to give

$$
M_1(h) = n^{-2} h^{-1} \sum_i \sum_j K^*\{h^{-1}(X_i - X_j)\} + 2n^{-1} h^{-1} K(0)
\tag{3.39}
$$

where the function K^* is defined by

$$
K^*(t) = K^{(2)}(t) - 2K(t).
\tag{3.40}
$$

More remarks about computational aspects will be made in Section 3.5. We shall see there that a Fourier series idea makes it possible to compute $M_1(h)$ very economically indeed.

A very strong large-sample justification of least-squares cross-validation is given by a remarkable result of Stone (1984). Given a sample, $X_1, \ldots, X_n$ from a density f, let $I_{lsxv}(X_1, \ldots, X_n)$ be the integrated square error of the density estimate constructed using the smoothing parameter that minimizes the score function $M_1(h)$ over all $h \geqslant 0$. Let $I_{opt}(X_1, \ldots, X_n)$ be the integrated square error if h is chosen optimally for this sample, in other words the minimum value of $\int (\hat{f} - f)^2$ over all h, keeping the data fixed. Assuming only that f is bounded, and under very mild conditions on the kernel, Stone has shown that, with probability one,

$$\frac{I_{\text{lsxv}}(X_1,\ldots,X_n)}{I_{\text{opt}}(X_1,\ldots,X_n)} \to 1 \qquad \text{as } n \to \infty.$$

Thus, asymptotically, least-squares cross-validation achieves the best possible choice of smoothing parameter, in the sense of minimizing the integrated square error. Put another way, Stone's theorem says that the score function $M_1(h)$ tells us, asymptotically, just as much about the optimal smoothing parameter, from the integrated square error point of view, as if we actually knew the underlying density f.

Discretization errors in cross-validation

Despite the strong attractiveness of least-squares cross-validation, it is important to note that it may give poor results if applied without modification to discretized data. Real data are nearly always rounded or discretized to a greater or lesser degree, and large data sets are very often presented in the form of a very fine histogram. Since discretizing to a fine grid does not move any of the data points very far, we would hope and expect that applying our algorithms to the rounded data would give good results. Unfortunately, this may well not be the case.

In a discretized data set $X_1,\ldots,X_n$, let m be the number of pairs $i < j$ for which $X_i = X_j$. For example, if the data set is a histogram of counts k_r, then

$$m = \sum_r \tfrac{1}{2}k_r(k_r - 1). \tag{3.41}$$

If a data set of size n is discretized to a grid of k points, then, no matter how the data points fall, it can be shown by Jensen's inequality (see Feller, 1966, p. 152) that

$$\frac{m}{n} \geqslant \frac{1}{2}\frac{n}{k} - 1. \tag{3.42}$$

If the data are at all nonuniform, then m will generally be much larger than the minimum value given by (3.42). We shall see that, if m/n is larger than some threshold value β depending on the kernel K, then the least-squares cross-validation score $M_1(h)$ tends to minus infinity as h tends to zero, and hence least-squares cross-validation as it stands will choose the degenerate value $h = 0$ for the window width.

Define $\beta = \frac{1}{2} K^{(2)}(0)/\{2K(0) - K^{(2)}(0)\}$. The condition $\beta > 0$ is one of the regularity conditions required by Stone (1984) in his proof of the asymptotic efficiency of least-squares cross-validation. It will certainly hold for any kernel whose maximum value is $K(0)$. For the normal kernel $\beta = 1/(2\sqrt{2} - 1) = 0.55$.

Suppose that $m > \beta n$. Then

$$hM_1(h) = n^{-2} \sum_i \sum_j K^*\{h^{-1}(X_i - X_j)\} + 2n^{-1}K(0)$$

As h tends to zero, $K^*\{h^{-1}(X_i - X_j)\}$ will tend to zero if $X_i \neq X_j$, but will remain fixed at $K^*(0)$ if $X_i = X_j$. The number of terms in the summation for which $X_i = X_j$ is $2m + n$; hence

$$\lim_{h \to 0} hM_1(h) = n^{-2}(2m + n)K^*(0) + 2n^{-1}K(0)$$

$$= 2mn^{-2}K^*(0) + n^{-1}K^{(2)}(0)$$

$$= 2n^{-1}K^{(2)}(0)\{-mn^{-1}\beta^{-1} + 1\} < 0$$

since $mn^{-1} > \beta$ by hypothesis. Hence $M_1(h) \to -\infty$ as $h \to 0$, and hence least-squares cross-validation will lead to the degenerate choice $h = 0$ of smoothing parameter.

The message of this discussion is not just that it is dangerous to use least-squares cross-validation for discretized data. Its real importance is in demonstrating that the behaviour of $M_1(h)$ for small h is highly sensitive to very fine small-scale effects in the data. Therefore it would be wise only to minimize the score function over a range of values of h suggested by the discussion in Section 3.4.2. For example, one might search between $0.25h^*$ and $1.5h^*$, where h^* is the value given by (3.31).

3.4.4 Likelihood cross-validation

The method of likelihood cross-validation is a natural development of the idea of using likelihood to judge the adequacy of fit of a statistical model. It is of general applicability, not just in density estimation; see, for example Stone (1974) and Geisser (1975).

The rationale behind the method, as applied to density estimation, is the following. Suppose that, in addition to the original data set, an independent observation Y from f were available. Then the log likelihood of f as the density underlying the observation Y would be log $f(Y)$; regarding $\hat{f}$ as a parametric family of densities depending on the window width h, but with the data $X_1, \ldots, X_n$ fixed, would give log $\hat{f}(Y)$, regarded as a function of h, as the log likelihood of the smoothing parameter h.

Now, since an independent observation Y is not available, we could omit one of the original observations X_i from the sample used to construct the density estimate, and then use X_i as the observation Y. This would give log likelihood log $\hat{f}_{-i}(X_i)$, where $\hat{f}_{-i}$ is as defined in (3.34) above. Since there is nothing special about the choice of which observation to leave out, the log likelihood is averaged over each choice of omitted X_i, to give the score function

$$CV(h) = n^{-1} \sum_{i=1}^{n} \log \hat{f}_{-i}(X_i). \tag{3.43}$$

The likelihood cross-validation choice of h is then the value of h which maximizes the function $CV(h)$ for the given data.

The score function $CV(h)$ was suggested by Duin (1976) and by Habbema, Hermans and van der Broek (1974). It has strong intuitive appeal and does not present severe computational difficulties. An argument similar to that given in Section 3.4.3 shows that, speaking heuristically, maximizing $CV(h)$ should yield a density estimate which is close to the true density in terms of the Kullback–Leibler information distance, defined by

$$I(f, \hat{f}) = \int f(x) \log \{f(x)/\hat{f}(x)\} \, dx.$$

Working heuristically, we have, letting $\hat{f}_{n-1}$ be the estimate based on only $(n-1)$ observations,

$$E\{CV(h)\} = E \log \hat{f}_{-n}(X_n) = E \int f(x) \log \hat{f}_{n-1}(x) \, dx$$

$$\approx E \int f(x) \log \hat{f}(x) \, dx = -E\{I(f, \hat{f})\} + \int f \log f,$$

so that $-CV(h)$ is, up to a constant, an unbiased estimator of the expected Kullback–Leibler error for an estimate using the same window width on a sample of size $(n-1)$. Unfortunately, this argument can only be made watertight under rather strong assumptions on f. If, for example, f has unbounded support and the kernel function K has bounded support, then $I(f, \hat{f})$ will be $-\infty$ for all h.

It has been noted (see Scott and Factor, 1981) that the performance of $CV(h)$ is very sensitive to outliers. To gain some intuition as to why this might be so, consider what will happen if the support of K is restricted to $(-1, 1)$ and one of the observations, say X_1, is distance R away from all the others. Then, if $h < R$, $\hat{f}_{-1}(X_1)$ will be zero, and so $CV(h)$ will be $-\infty$ for all $h < R$, regardless of the behaviour of any of the other observations $X_2, \ldots, X_n$. Thus the maximizer of $CV(h)$ is forced to be greater than R, and this may well lead to oversmoothing. Even though this obvious difficulty may be circumvented by using a kernel with heavier tails, it is nevertheless worrying that the tails of the sample will still exert an undue influence because of the divergence of $\log f$ to $-\infty$ as $|x|$ gets large. Chow, Geman and Wu (1983) conjecture that the use of a kernel with heavy tails may actually lead to undersmoothed densities rather than the oversmoothing that may arise if the support of the kernel is bounded.

It is not just outliers that can raise difficulties if likelihood cross-validation is used to choose the smoothing parameter. One of the mildest requirements of a statistical procedure is that it should be consistent, that is to say that, speaking loosely, good estimates of the quantity of interest should be obtained if very large samples are available. Under the assumption that the true density f is uniformly continuous, we shall see in Section 3.7 that, for all t, $\hat{f}(t)$ is consistent as an estimator of $f(t)$ under very mild conditions on the large-sample behaviour of the window width h; one of these conditions is that $h \to 0$ as $n \to \infty$.

Now suppose that one or other of the tails of f is eventually monotonic and dies off at an exponential rate, or more slowly. This is the case for virtually all standard probability densities, except for the normal and those of bounded support. Schuster and Gregory (1981) show that in this case, which is very far from pathological, the use of likelihood cross-validation actually leads to inconsistent estimates of the density. Their argument rests on the property that, under the given assumptions, the gaps between extreme observations in the tails will not become smaller as the sample size increases. Therefore, by the argument given above, the window width chosen by maximizing $CV(h)$ cannot converge to zero as n tends to infinity.

When applied to discretized data, likelihood cross-validation also suffers – but not to the same extent – from the difficulties that occur when using least-squares cross-validation. The score function $CV(h)$ will certainly tend to infinity as $h \to 0$ if there are no isolated data points at all, by no means an exceptional occurrence in large discretized data sets. If there are some isolated data points but a large number of coincidences in the data, then the limiting behaviour of $CV(h)$ as $h \to 0$ will depend on the tail properties of the kernel; further details are left to the reader to investigate. Suffice it to say that, as with least-squares cross-validation, it would be wise to narrow down the range of possible h somewhat before blindly applying likelihood cross-validation.

3.4.5 *The test graph method*

The test graph method, developed by Silverman (1978a), is an entirely different approach from the two cross-validation methods discussed above. It aims to yield estimates that are uniformly close to the true density. On a finite interval at least, this is a somewhat stronger requirement than small integrated square error, since convergence of $\sup |\hat{f} - f|$ to zero is sufficient, but not necessary, for convergence of $\int (\hat{f} - f)^2$ to zero. The method is partly subjective as it stands. It is of interest both in its own right and because of the intuition it gives about the way that derivatives of the density estimate will behave if the estimate itself is close to the true density.

The rationale behind the method is a theorem, stated and proved by Silverman (1978a), which gives the following result. Suppose the kernel K is symmetric and twice differentiable and satisfies certain regularity conditions, and that $\int x^2 K(x) \, dx$ is nonzero. Suppose also that the unknown density f has uniformly continuous and bounded second derivative. Now suppose that h is chosen, as a function of n, to ensure the most rapid possible convergence of $\sup |\hat{f} - f|$ to zero; in other words, h is chosen to minimize the maximum error in the estimation of the density. Then, using the same choice of window width it will be the case that, as $n \to \infty$,

$$\frac{\sup |\hat{f}'' - E\hat{f}''|}{\sup |E\hat{f}''|} \to k \qquad (3.44)$$

where the constant k depends only on the kernel and is given by

$$k = \tfrac{1}{2} \int |x^2 K(x) \, dx| \left\{ \int (K'')^2 \, dx \Big/ \int K^2 \, dx \right\}^{1/2}.$$

If K is the Gaussian kernel then the constant k is approximately 0.4. The term $\hat{f}'' - E\hat{f}''$ in the numerator of (3.44) represents the random noise in the curve $\hat{f}''$, while the denominator term $E\hat{f}''$ is the trend of this curve. Thus (3.44) can be restated in words as saying that, for good estimation of the density itself, the magnitude of the noise in $\hat{f}''$ will be about half of the maximum value of the trend of this curve. For reasonably large sample sizes, the noise will appear as rapid fluctuations in the curve $\hat{f}''$.

The proposed method for choosing the smoothing parameter is as follows. Draw 'test graphs' of the second derivative of $\hat{f}$ for various values of h. In the light of the discussion above, the ideal test graph should have rapid fluctuations which are quite marked but do not obscure the systematic variation completely. Choose the window width which yields a test graph conforming to this principle, and use this window width for estimating the density itself.

An example of the application of the test graph principle is given in Fig. 3.6. Here the underlying data consist of 300 independent simulated realizations of a test statistic, the maximum of the cosine quantogram, which is of interest in archaeological statistics. The data were obtained and kindly provided by John Kent; for further details see Silverman (1978a) and Kent (1976). Test graphs for the three window widths 2.5, 2.9 and 3.3 are illustrated in Fig. 3.6. It can be seen that, as the window width increases over the quite narrow range

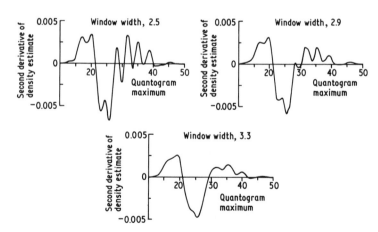

Fig. 3.6 *Test graphs for cosine quantogram data with window widths 2.5, 2.9 and 3.3. Reproduced from Silverman (1978a) with the permission of the Biometrika Trustees.*

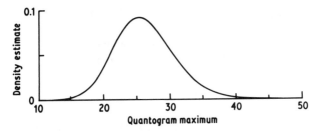

Fig. 3.7 *Density estimate for cosine quantogram data, window width 2.9. Reproduced from Silverman (1978a) with the permission of the Biometrika Trustees.*

considered, the test graph is 'breaking up' and that rapid local fluctuations are becoming clearly visible. If window widths much below 2.5 are considered, the test graph becomes very noisy indeed, while window widths above 3.3 give smooth curves with little or no visible random noise. On the basis of these curves, the test graph principle stated above suggests that a window width of around 2.9 is appropriate for the estimation of f itself, and the corresponding estimate is given in Fig. 3.7. The skewness of the distribution is clearly visible.

In practice the character of the test graph changes much more rapidly than that of the density estimate itself, and so the choice of an appropriate test graph is by no means as difficult, or as critical, as might at first be supposed. Some further examples are given in Silverman (1978a) and a tentative proposal for the objective assessment of test graphs is discussed in Silverman (1980). Because of the degree of subjectivity involved, and also the dependence in the justification on rather sensitive asymptotic results, the test graph method's main appeal to many readers may well be as a check on other methods of smoothing parameter choice.

3.4.6 *Internal estimation of the density roughness*

Consider the equation (3.21) for the optimal window width. If we define $\alpha(K) = k_2^{-2/5}\{\int K(t)^2 \, dt\}^{1/5}$ and $\beta(f) = (\int f''^2)^{-1/5}$, this equation becomes

$$h_{\text{opt}} = \alpha(K)\beta(f)n^{-1/5}. \tag{3.45}$$

A natural approach based on (3.45) would be to use an initial window width h_0 to provide an estimate $\hat{\beta}(h_0)$ of $\beta(f)$; this estimate would be substituted back into (3.45) to give the h actually used for the estimation of the density. The estimate of $\beta(f)$ would be given by

$$\hat{\beta}(h_0) = \left(\int \hat{f}_0''^2 \right)^{-1/5} = \beta(\hat{f}_0)$$

where $\hat{f}_0$ is the density estimate constructed from the data with window width h_0. If the kernel K is the standard normal density, then (see Scott and Factor, 1981, equation 2.11) elementary but tedious manipulation leads to

$$\hat{\beta}(h)^{-5} = \tfrac{3}{8} \pi^{-1/2} n^{-2} h^{-5} \sum_{j=1}^{n} \sum_{k=1}^{n} \Psi\{(X_j - X_k)/h\}$$

where $\Psi(u) = (1 - u^2 + u^4/12) \exp(-\tfrac{1}{4} u^2)$.

The window width actually used would then be h_1, where

$$h_1 = \alpha(K) \hat{\beta}(h_0) n^{-1/5}. \tag{3.46}$$

The approach of estimating the quantity $\beta(f)$ from an initial density estimate was suggested, essentially, by Woodroofe (1970). His formulation was based on the mean square error in the estimation of the density at a single point, rather than the mean integrated square error, but the basic idea is the same. He also provided some theoretical justification for his procedure. An obvious drawback is that it is still necessary to choose an initial h_0; Woodroofe suggests that the choice of h_0 is somewhat less sensitive than the direct choice of the window width to be used for the estimation of f. However, it is intuitively clear that the method will be self-serving to some extent. A large value of h_0 will lead to a smoother $\hat{f}_0$, and hence a relatively larger value of $\hat{\beta}(h_0)$ and of h_1.

Scott and Factor (1981) give a plot, given as Fig. 3.8, which demonstrates the relationship between h_0 and h_1 for a particular data set (given as Table 3 in their paper). Here the data consist of 686 values, discretized to a fine grid, of a variable observed by a remote sensing satellite. The fairly strong dependence between h_1 and h_0 is clear from Fig. 3.8. In particular if h_0 is fairly small, h_1 will be almost exactly equal to h_0.

In order to avoid the problem of choosing an initial value h_0, Scott, Tapia and Thompson (1977) suggested an iterative approach where,

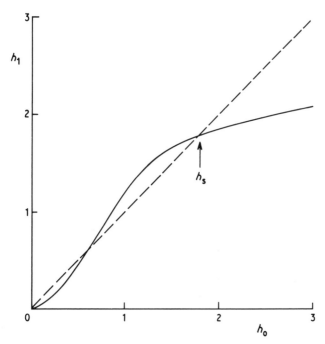

Fig. 3.8 *Relationship (3.46) between h_0 and h_1, for remote sensing brightness data. Adapted from Scott and Factor (1981) with the permission of the American Statistical Association.*

starting with a large h_0, values $h_1, h_2, \ldots$ are found satisfying

$$h_i = \alpha(K)\hat{\beta}(h_{i-1})n^{-1/5} \qquad (3.47)$$

The iteration is continued to convergence; this amounts to choosing h to be the largest solution of the equation

$$h = \alpha(K)\hat{\beta}(h)n^{-1/5} \qquad (3.48)$$

Call this solution h_s. In the remote sensing example it can be seen from Fig. 3.8 that this procedure would lead to a choice $h_s = 2.0$. In practice it is faster to solve (3.48) by Newton's method rather than by iteration; occasionally the only root of (3.48) will be the degenerate solution $h = 0$, but usually there will be at least one strictly positive root.

The density estimate for the remote sensing data with window width 2.0 is given in Fig. 3.9. It is interesting to note that the simple

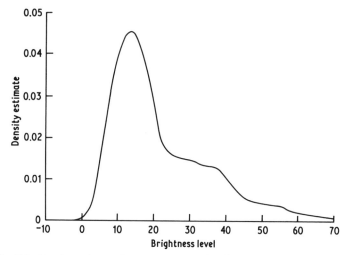

Fig. 3.9 *Density estimate for remote sensing brightness data, window width h_s chosen from Fig. 3.8 by Scott–Tapia–Thompson method. Reproduced from Scott and Factor (1981) with the permission of the American Statistical Association.*

rule (3.31) yields the somewhat larger window width $h = 3.3$ for these data.

Some slightly disturbing intuition about the behaviour of the Scott–Tapia–Thompson procedure may be gained by considering the theory of the test graph procedure discussed in Section 3.4.5. Speaking heuristically, a good h for the estimation of the density will not yield a very good estimate of the second derivative, and hence it cannot be expected that $\beta(f)$ will be very well estimated by $\hat{\beta}(h)$. Nevertheless, simulation studies reported by Scott, Tapia and Thompson (1977) and Scott and Factor (1981) show that h_s is a reasonably good choice of smoothing parameter under a variety of models. The latter study includes a comparison with likelihood cross-validation, and indicates that for short-tailed densities, likelihood cross-validation performs slightly better. However, the Scott–Tapia–Thompson method is insensitive to outliers, in sharp contrast to likelihood cross-validation.

Bowman (1985) reports a careful simulation study in which several of the methods described in this section were applied to samples of sizes 25, 50 and 100 drawn from a variety of distributions, including bimodal, skew and heavy-tailed densities. In this study, the performance of the Scott–Tapia–Thompson method was disappointing. Likelihood cross-validation was quite successful for short-tailed distributions but (as would be expected) gave poor results for densities with heavy tails. Particularly for the larger sample sizes, least-squares cross-validation performed quite well overall. A rule of thumb similar to (3.31) gave even better results than least-squares cross-validation for unimodal densities, especially for smaller sample sizes, but (again not surprisingly) oversmoothed bimodal densities somewhat. As was pointed out in Section 3.4.2, at least in the case of equal mixtures of normal densities, the degree by which (3.31) oversmooths bimodal densities will not usually be sufficient to obscure the bimodality completely, provided samples of reasonable size are available.

3.5 Computational considerations

At first sight it may appear surprising that the first published paper on kernel density estimation appeared as late as 1956, considering that the basic idea is, in retrospect, so simple. Of course, analogous remarks could be made about most scientific discoveries! It should be stressed, though, that before the widespread availability of computers, the calculation and plotting of a kernel density estimate from a given data set would have been a formidable task. Even with present-day computational power, it is all too easy to consume inordinate amounts of computer time by using inefficient algorithms for finding density estimates. The use of an inappropriate algorithm may not be too disastrous when plotting out a couple of estimates based on moderate sized data sets, but the difficulties become more serious when very large samples are involved, or if a substantial number of estimates are to be calculated. Furthermore, the use of techniques like least-squares cross-validation can easily consume large amounts of computer time because it appears from the formula (3.39) that $\frac{1}{2}n(n-1)$ evaluations of the function K^* are needed to compute each value of the score function $M_1(h)$; since it is then necessary to minimize over h, the computations could easily get out of hand.

For all but very small samples, the direct use for computation of the defining formula for the kernel estimate is highly inefficient. It is much faster to notice that the kernel estimate is a convolution of the data with the kernel and to use Fourier transforms to perform the convolution. The use of the fast Fourier transform (see, for example,

Monro, 1976) makes it possible to find direct and inverse Fourier transforms very quickly. Furthermore, we shall see below that the least-squares cross-validation score can be found remarkably quickly by Fourier transform methods. The algorithms developed in this section are an enhanced version of Silverman (1982a). See also the improvements to that algorithm suggested by Jones and Lotwick (1984).

Given any function g, denote by $\tilde{g}$ its Fourier transform

$$\tilde{g}(s) = (2\pi)^{-1/2} \int e^{ist} g(t)\, dt.$$

Define $u(s)$ to be the Fourier transform of the data,

$$u(s) = (2\pi)^{-1/2} n^{-1} \sum_{j=1}^{n} \exp(isX_j).$$

Let $\tilde{f}_n(s)$ be the Fourier transform of the kernel density estimate; taking Fourier transforms of the definition of the kernel density estimate yields

$$\tilde{f}_n(s) = (2\pi)^{1/2} \tilde{K}(hs) u(s) \tag{3.49}$$

by the standard convolution formula for Fourier transforms. Here we have used the property that the Fourier transform of the scaled kernel $h^{-1}K(h^{-1}t)$ is $\tilde{K}(hs)$. Formula (3.49) is particularly suitable for use when K is the Gaussian kernel; in this case the Fourier transform of K can be substituted explicitly to yield

$$\tilde{f}_n(s) = \exp(-\tfrac{1}{2}h^2 s^2) u(s). \tag{3.50}$$

The basic idea of the algorithm we shall develop in this section is to use the fast Fourier transform both to find the function u and also to invert $\tilde{f}_n$ to find the density estimate $\hat{f}$.

Finding the least-squares cross-validation score $M_1(h)$ from the Fourier transform is straightforward. Define

$$v(s) = (2\pi)^{-1/2} n^{-2} \sum\sum \exp\{is(X_j - X_k)\}$$
$$= (2\pi)^{1/2} |u(s)|^2 \tag{3.51}$$

The Fourier transform of the function K^* defined in (3.40) is

$$\tilde{K}^*(s) = \tilde{K}^{(2)}(s) - 2\tilde{K}(s)$$
$$= (2\pi)^{1/2} \tilde{K}(s)^2 - 2\tilde{K}(s) \tag{3.52}$$
$$= (2\pi)^{-1/2} \{\exp(-s^2) - 2\exp(-\tfrac{1}{2}s^2)\} \tag{3.53}$$

in the special case of the Gaussian kernel. Define a function

$$\psi(t) = n^{-2} \sum_i \sum_j h^{-1} K^* \{(X_i - X_j)h^{-1} - t\} \qquad (3.54)$$

Then the least-squares cross-validation score is given by

$$M_1(h) = \psi(0) + 2n^{-1}h^{-1}K(0). \qquad (3.55)$$

Now

$$\begin{aligned}
\tilde{\psi}(s) &= (2\pi)^{1/2}\tilde{K}^*(hs)v(s) \\
&= 2\pi\tilde{K}^*(hs)|u(s)|^2
\end{aligned}$$

and so

$$\begin{aligned}
\psi(0) &= (2\pi)^{-1/2}\int \tilde{\psi}(s)\,ds \\
&= (2\pi)^{1/2}\int \tilde{K}^*(hs)|u(s)|^2\,ds \\
&= \int \{\exp(-h^2s^2) - 2\exp(-\tfrac{1}{2}h^2s^2)\}|u(s)|^2\,ds
\end{aligned}$$

$$(3.56)$$

if the Gaussian kernel is used. Substituting (3.56) into (3.55) gives the least-squares cross-validation score; note that it is not even necessary to invert the transform to find this score.

Since the fast Fourier transform gives the discrete Fourier transform of a sequence rather than the Fourier transform of a function, it is necessary to make some slight adjustment to the procedure. Consider an interval $[a, b]$ on which all the data points lie. The effect of the Fourier transform method we shall describe is to impose periodic boundary conditions, by identifying the end points a and b, and so the interval should be chosen large enough that this does not present any difficulties. Putting $a < \min(X_i) - 3h$ and $b > \max(X_i) + 3h$ is ample for this purpose if the Gaussian kernel is being used. Of course, if periodic boundary conditions on a given interval are actually required, for example in the case of directional data, then that interval should be used as the interval $[a, b]$.

Choose $M = 2^r$ for some integer r; the density estimate will be found at M points in the interval $[a, b]$ and choosing $r = 7$ or 8 will give excellent results. Define

$$\begin{aligned}
\delta &= (b - a)/M \\
t_k &= a + k\delta \qquad \text{for } k = 0, 1, \ldots, M - 1.
\end{aligned}$$

Discretize the data as follows. If a data point X falls in the interval $[t_k, t_{k+1}]$, it is split into a weight $n^{-1}\delta^{-2}(t_{k+1} - X)$ at t_k and a weight

$n^{-1}\delta^{-2}(X-t_k)$ at t_{k+1}; these weights are accumulated over all the data points X_i to give a sequence (ξ_k) of weights summing to δ^{-1}. Now, for $-\frac{1}{2}M \leqslant l \leqslant \frac{1}{2}M$, define Y_l to be the discrete Fourier transform

$$Y_l = M^{-1} \sum_{k=0}^{M-1} \xi_k \exp\{i2\pi kl/M\}$$

which may be found by fast Fourier transformation.
Define

$$s_l = 2\pi l(b-a)^{-1}$$

and assume for the moment that $a = 0$. Then, using the definition of the weights ξ_k,

$$Y_l = M^{-1} \sum_k \xi_k \exp\{it_k s_l\}$$

$$\approx n^{-1}M^{-1}\delta^{-1} \sum_j \exp\{is_l X_j\} \qquad (3.57)$$

$$= (2\pi)^{1/2}(b-a)^{-1}u(s_l). \qquad (3.58)$$

The approximation (3.58) will deteriorate as $|s_l|$ increases, but, since the next step in the algorithm multiplies all Y_l for larger $|l|$ by a very small factor, this will not matter in practice.
Define a sequence ζ_l^* by

$$\zeta_l^* = \exp(-\tfrac{1}{2}h^2 s_l^2)Y_l \qquad (3.59)$$

and let ζ_k be the inverse discrete Fourier transform of ζ_l^*. Then

$$\zeta_k = \sum_{l=-M/2}^{M/2} \exp(-2\pi ikl/M)\zeta_l^*$$

$$\approx \sum_l \exp(-is_l t_k)(2\pi)^{1/2}(b-a)^{-1}\exp(-\tfrac{1}{2}h^2 s_l^2)u(s_l)$$

$$\approx (2\pi)^{-1/2}\int \exp(-ist_k)\exp(-\tfrac{1}{2}h^2 s^2)u(s)\,ds \qquad (3.60)$$

$$= \hat{f}(t_k)$$

since (3.60) is the inverse Fourier transform of $\hat{f}$ as derived in (3.49). The case of general a yields algebra which is slightly more complicated, but the end result (3.59) is exactly the same. A detailed analysis of the various errors introduced in this algorithm is given by Jones and

Lotwick (1983). For almost all practical purposes the errors are negligible.

Thus the density estimate can be found on the lattice t_k by the following algorithm:

Step 1 Discretize to find the weight sequence ξ_k.
Step 2 Fast Fourier transform to find the sequence Y_l.
Step 3 Use (3.59) to find the sequence ζ_l^*.
Step 4 Inverse fast Fourier transform to find the sequence $\hat{f}(t_k)$.
Step 5 If estimates with other window widths are required for the same data, repeat steps 3 and 4 only.

Considerable economies are available, both in storage and computer time, by using the fact that (Y_l) is the Fourier transform of a real sequence; see Monro (1976) for an algorithm that takes advantage of this property.

Turn now to the question of finding the least-squares cross-validation score. We have, approximating the integral (3.56) by a sum, and substituting (3.58),

$$
\psi(0) = (b-a) \sum_{l=-M/2}^{M/2} \{\exp(-h^2 s_l^2) - 2\exp(-\tfrac{1}{2}h^2 s_l^2)\} |Y_l|^2
$$
$$
= -1 + 2(b-a) \sum_{l=1}^{M/2} \{\exp(-h^2 s_l^2) - 2\exp(-\tfrac{1}{2}h^2 s_l^2)\} |Y_l|^2
$$
$$
\tag{3.61}
$$

since $Y_0 = M^{-1}\Sigma \xi_k = M^{-1}\delta^{-1} = (b-a)^{-1}$ and $|Y_l| = |Y_{-l}|$ for all l. Substituting (3.61) back into (3.55) gives

$$
\tfrac{1}{2}\{1 + M_1(h)\} = (b-a) \sum_{l=1}^{M/2} \{\exp(-h^2 s_l^2)
$$
$$
- 2\exp(-\tfrac{1}{2}h^2 s_l^2)\} |Y_l|^2 + n^{-1}h^{-1}(2\pi)^{-1/2}. \tag{3.62}
$$

This criterion is easily found for a range of values of h. For the values that will be of interest, the exponential terms will rapidly become negligible and so the sum actually calculated will have far fewer than $\tfrac{1}{2}M$ terms. Very little detailed work has been done on the form of the score function $M_1(h)$ and so strong recommendations cannot be made about an appropriate minimization algorithm for $M_1(h)$. However, reference to Section 3.4.2 will give an indication of the likely interval in which to search for the minimum of M_1; one could start by searching, for instance, for h in the interval

$$\tfrac{1}{4}n^{-1/5}\sigma < h < \tfrac{3}{2}n^{-1/5}\sigma \qquad (3.63)$$

and then extend the interval if the minimum occurs at the edge of this range. Since, up to a constant, M_1 is asymptotically a good estimate of the integrated square error, and since the approximate form (3.20) for the mean integrated square error is convex in h, it would be a little surprising to find severe nonconvexity in M_1. A rather conservative minimization strategy would be to perform a grid search in the interval (3.63) and then improve the best point found by a quasi-Newton approach.

3.6 A possible bias reduction technique

So far in this chapter we have concentrated attention on the case of kernels that satisfy the conditions (3.12), usually being symmetric probability density functions. There are some arguments in favour of using kernels which take negative as well as positive values, and these will be discussed in this section. These arguments were first put forward by Parzen (1962) and Bartlett (1963).

3.6.1 Asymptotic arguments

Suppose we relax the condition that K is to be non-negative and choose as kernel a symmetric function K that satisfies

$$\int K(t)\,dt = 1, \quad \int t^2 K(t)\,dt = 0 \quad \text{and} \quad \int t^4 K(t)\,dt = k_4 \neq 0 \quad (3.64)$$

Notice that the conditions (3.64) cannot be satisfied if K is a non-negative kernel, because it would then be impossible for $\int t^2 K(t)\,dt$ to be zero.

Similar arguments to those in Section 3.3.1 can be used to obtain an approximation for the bias of the density estimate. Pursuing the Taylor expansion of $f(x - ht)$ to terms of order h^4, instead of just h^2 as before, gives

$$f(x - ht) = f(x) - ht f'(x) + \tfrac{1}{2}h^2 t^2 f''(x) \\ - \tfrac{1}{6}h^3 t^3 f'''(x) + \tfrac{1}{24}h^4 t^4 f^{\text{iv}}(x) + \cdots \qquad (3.65)$$

Substituting (3.65) into the expression for $bias_h(x)$ as in equation (3.15) then yields an expansion in which the terms in h, h^2 and h^3 are all zero. The terms in h and h^3 drop out because of the symmetry of K, while the term in h^2 is zero because of condition (3.64), which says that the coefficient k_2 of (3.16) is zero. The expression corresponding to (3.16) is now

$$bias_h(x) = \tfrac{1}{24} h^4 f^{iv}(x) k_4 + \text{higher-order terms in } h. \qquad (3.66)$$

Thus the use of a kernel satisfying (3.64) has reduced the approximate bias from order h^2 to order h^4.

The calculation of the approximate expression (3.18) for the variance goes through exactly as before. Just as in Section 3.3.2, the approximations for the bias and variance can be used to obtain an approximate expression for the mean integrated square error:

$$\tfrac{1}{576} h^8 k_4^{\,2} \int f^{iv}(x)^2 \, dx + n^{-1} h^{-1} \int K(t)^2 \, dt. \qquad (3.67)$$

Minimizing (3.67) over h gives as the approximately optimal value for h:

$$h_{opt} = (72)^{1/9} k_4^{-2/9} \left\{ \int K(t)^2 \, dt \right\}^{1/9} \left\{ \int f^{iv}(x)^2 \, dx \right\}^{-1/9} n^{-1/9}. \qquad (3.68)$$

Finally, this value of the window width can be substituted back into (3.67) to give as the minimum achievable mean square error the value

$$C_4(K) \left\{ \int f^{iv}(x)^2 \, dx \right\}^{1/9} n^{-8/9} \qquad (3.69)$$

where the constant $C_4(K)$ is given by

$$C_4(K) = 9^{8/9} 2^{-10/3} k_4^{\,2/9} \left\{ \int K(t)^2 \, dt \right\}^{8/9}. \qquad (3.70)$$

The basic message of these asymptotic calculations is that the main advantage of using a kernel satisfying (3.64) is a small improvement in the order of magnitude of the best achievable mean integrated square error, from $n^{-4/5}$ to $n^{-8/9}$. For all except astronomical sample sizes, any case for using such kernels, on the basis of mean integrated square error, must rest on a careful comparison of the constants by which these factors are multiplied. Furthermore, the calculations leading to

(3.66) and the subsequent expressions are extremely delicate, because they depend both on the high-order smoothness properties of the unknown density f and on a long Taylor expansion.

3.6.2 Choice of kernel

A discussion of the choice of possible kernels satisfying (3.64) is given by Deheuvels (1977) and Müller (1984). The natural approach would be to attempt to minimize the quantity $C_4(K)$ subject to the constraints that K satisfied (3.64) and integrated to one; however this minimization problem has no solution since $C_4(K)$ can be made arbitrarily small even subject to the constraints. An alternative approach is to minimize the asymptotic variance of the density estimate subject to these constraints and in addition the constraint that K is of bounded support. This leads to the kernel

$$K(y) = \begin{cases} \frac{3}{8}(3 - 5y^2) & |y| < 1 \\ 0 & \text{otherwise.} \end{cases} \tag{3.71}$$

This kernel is discontinuous at ± 1 and so leads to estimates that are themselves discontinuous, an unfortunate property in view of the theory leading to the use of such kernels, which requires the unknown density f to be four times differentiable.

Schucany and Sommers (1977) suggest an attractive 'jackknife' method that implicitly constructs a kernel satisfying (3.64). Given a sample $X_1, \ldots, X_n$, let f_h be the estimate with window width h constructed from the sample using a positive kernel K_0, such as the normal density function. The $O(h^2)$ term in the bias of f_h is eliminated by constructing the estimator

$$\hat{f}(t) = \frac{f_h(t) - c^{-2} f_{ch}(t)}{1 - c^{-2}}, \tag{3.72}$$

a linear combination of two estimators with different bandwidths. The constant c is fixed by the user. If the Fourier transform technique described in Section 3.5 is used to calculate f_h, then finding a second estimator f_{ch} based on the same data, and hence evaluating the $\hat{f}$ of (3.72), is straightforward.

The estimator $\hat{f}$ of (3.72) can easily be seen to be a kernel estimator with kernel K given by

$$K(t) = \frac{K_0(t) - c^{-3} K_0(c^{-1}t)}{1 - c^{-2}} \tag{3.73}$$

and it is easy to show that this kernel satisfies the conditions (3.64).

Schucany and Sommers suggest using a value of c near 1, but detailed work on an appropriate choice of c remains to be done.

The limit of (3.73) as c tends to one is the kernel

$$K_1(t) = \tfrac{3}{2} K_0(t) + \tfrac{1}{2} t K_0{}'(t); \tag{3.74}$$

this formula is derived by an application of Taylor's theorem. If K_0 is the normal density ϕ, then (3.74) becomes

$$K_1(t) = (\tfrac{3}{2} - \tfrac{1}{2} t^2)\, \phi(t) = \phi(t) - \tfrac{1}{2}\, \phi''(t) \tag{3.75}$$

and the Fourier transform methods of Section 3.5 can be adapted easily to find the kernel estimate with kernel K_1. The Fourier transform of K_1 will be

$$K_1{}^*(s) = \phi^*(s) + \tfrac{1}{2} s^2 \phi^*(s), \tag{3.76}$$

so that (3.50) will be replaced by

$$\tilde{f}_n(s) = (1 + \tfrac{1}{2} h^2 s^2) \exp\left(- \tfrac{1}{2} h^2 s^2\right) u(s). \tag{3.77}$$

3.6.3 Discussion

Very little practical work has been done on the evaluation of the performance of kernels satisfying (3.64). Since such kernels inevitably take negative values and have subsidiary local maxima, the estimates obtained using them may well exhibit similar behaviour. The 'spline transform' technique of Boneva, Kendall and Stefanov (1971, Section 3) is, fortuitously, a kernel density estimation method using a kernel satisfying (3.64). The authors of that paper remark (p. 35) that the sign-varying nature of their kernel can lead to fluctuations in the density estimate; they point out, with an example, that these can occasionally be turned to good account when it is of interest to detect discontinuities in the density f.

Schucany and Sommers (1977) report a simulation study in which their technique performs well at estimating the value of a density at one (rather special) point. However, if the interest is in discerning the general shape of the density, it may be better to tolerate some bias rather than risk the introduction of spurious fluctuations in the tails and the possible exaggeration of such things as modes in the data. In some applications it is essential that the density estimate is itself a probability density function and so the use of a non-negative kernel is essential. Cautious users are best advised to stick to symmetric non-negative unimodal kernels. A method that has bias of the same order as (3.66), but still yields non-negative estimates, will be described in Section 5.3.3.

The discussion of this section can be extended to cover kernels that satisfy higher-order versions of (3.64). If we are prepared to assume that the unknown density has $2m$ derivatives, then there is an argument for using a kernel K satisfying

$$\int t^j K(t)\,dt = 0 \qquad \text{for } 0 < j < 2m, \qquad \int t^{2m} K(t)\,dt \neq 0 \qquad (3.78)$$

since, by similar manipulations to those conducted above, the bias in the estimation will then be of order h^{2m} and the optimal mean integrated square error of order $n^{-4m/(4m+1)}$. However, the difficulties mentioned above become more serious since the appropriate kernels are likely to have more subsidiary maxima and heavier negative parts. Furthermore, the asymptotic justification becomes still more delicate and depends on very high differentiability of the unknown underlying density. Müller (1984) lists kernels which have the required properties (3.78) and are optimal in the same sense as the kernel given in (3.71) above.

3.7 Asymptotic properties

There is an enormous literature on the asymptotic properties of density estimation in general and on the kernel estimate in particular. In this section a few of the asymptotic results will be discussed, both to give the general flavour of what has been proved and to give some intuition about the large-sample behaviour of the estimates. The reader with a serious interest in the asymptotics of density estimation is referred for reviews of the literature to Prakasa Rao (1983), Wertz (1978), Wertz and Schneider (1979) and Rosenblatt (1971), and also to the various references cited below. Many of the techniques that have been used to prove the results discussed are extremely cunning and technical; no attempt will be made to describe them here.

The usual asymptotic framework in which theorems about kernel density estimation are proved is to assume that the kernel K and the unknown density f are fixed and satisfy given regularity conditions. The density estimates considered are constructed from the first n observations in an independent identically distributed sequence X_1, $X_2, \ldots$ drawn from f. It is assumed that the window width h depends in some way on the sample size n. Limiting results are then obtained on the behaviour of the estimate as n tends to infinity. In order to make the dependence on n explicit, we shall write h_n for the window width in this section.

3.7.1 Consistency results

Much attention has been paid to conditions under which the kernel estimate is, in various senses, a consistent estimate of the true density. The conditions for consistency are surprisingly mild, though the rate at which the estimated density converges to its true value can be extremely slow.

Consistency of the estimate f at a single point x was studied by Parzen (1962). His assumptions on the kernel K were that K was a bounded Borel function, satisfying

$$\int |K(t)|\,dt < \infty \qquad \text{and} \qquad \int K(t)\,dt = 1 \qquad (3.79)$$

and

$$|tK(t)| \to 0 \qquad \text{as } |t| \to \infty. \qquad (3.80)$$

These conditions are satisfied by almost any conceivable kernel.

The window width h_n was assumed to satisfy

$$h_n \to 0 \qquad \text{and } nh_n \to \infty \qquad \text{as } n \to \infty; \qquad (3.81)$$

under these conditions it was shown that, provided f is continuous at x,

$$\hat{f}(x) \to f(x) \text{ in probability as } n \to \infty.$$

The conditions (3.81) are typical of those required for consistency. They imply that, while the window width must get smaller as the sample size increases, it must not converge to zero as rapidly as n^{-1}. This is to say that the expected number of points in the sample falling in the interval $x \pm h_n$ must tend to infinity, however slowly, as n tends to infinity.

Once we move away from consistency at a single point, it becomes necessary to specify in what way the curve estimate $\hat{f}$ approximates the true density f. Uniform consistency – that is convergence in probability of $\sup |\hat{f}(x) - f(x)|$ to zero – has been considered by several authors, for example Parzen (1962), Nadaraya (1965), Silverman (1978b) and Bertrand-Retali (1978).

Suppose the kernel K is bounded, has bounded variation and satisfies (3.79), and that the set of discontinuities of K has Lebesgue measure zero. Again, these conditions are satisfied by virtually any conceivable kernel. Suppose that

$$f \text{ is uniformly continuous on } (-\infty, \infty) \qquad (3.82)$$

and that the window width h_n satisfies

$$h_n \to 0 \quad \text{and} \quad nh_n(\log n)^{-1} \to \infty \quad \text{as } n \to \infty. \qquad (3.83)$$

Bertrand-Retali (1978) shows, by an ingenious but complicated argument, that it will then be the case that, with probability 1,

$$\sup_x |\hat{f}(x) - f(x)| \to 0 \quad \text{as } n \to \infty$$

and furthermore that the conditions (3.82) and (3.83) are necessary as well as sufficient for uniform consistency. The conditions (3.83) are only very slightly stronger than those (3.81) required for pointwise consistency.

A different, and somewhat weaker, kind of consistency is discussed in depth by Devroye and Györfi (1985). Again, they use a complicated technical argument to obtain a most remarkable result. Assume that K is a non-negative Borel function which integrates to one. Making no assumptions whatsoever on the unknown density f, they show that conditions (3.81) are necessary and sufficient for the convergence

$$\int |\hat{f}(x) - f(x)| \, dx \to 0 \text{ with probability one as } n \to \infty. \qquad (3.84)$$

3.7.2 Rates of convergence

The extremely weak conditions (3.81) and (3.83) under which the estimates have been shown to be consistent might well trap the unwary into a false sense of security, by suggesting that good estimates of the density can be obtained for a wide range of values of the window width h. As we have already seen, this is far from being the case. The observed sensitivity of the estimates to the value of the window width can be reconciled with the result of Section 3.7.1 by considering the rate at which the estimate $\hat{f}$ converges to the true density f.

The approximations for the mean integrated square error developed in Section 3.3 are an example of a result giving the rate of convergence of $\hat{f}$ to f in a particular sense. They can be formalized into a rigorous theorem by treating the various Taylor expansions and integrals in Section 3.3.1 more carefully; see, for example, Prakasa Rao (1983, Theorem 2.1.7).

Consider the approximation (3.20) for the mean integrated square error for a moment. Recall that (3.22) showed that, if h is chosen optimally, under suitable regularity conditions the approximate value of the mean integrated square error will tend to zero at the rate $n^{-4/5}$. Suppose, now, that instead of choosing h optimally we set $h = n^{-1/2}$, a value that might be suggested by the consistency conditions (3.81). Substituting this value back into (3.20) shows that the mean integrated square error would then only tend to zero at rate $n^{-1/2}$.

Results can also be obtained for the rate of convergence of $\hat{f}$ to f in various senses other than mean integrated square error. For example, Silverman (1978a) obtained the exact rates at which sup $|\hat{f}(x) - E\hat{f}(x)|$ and sup $|E\hat{f}(x) - f(x)|$ converge to zero; these results were used in the development of the theory underlying the test graph method for choosing the window width, described in Section 3.4.5 above. Under rather restrictive conditions, Bickel and Rosenblatt (1973) studied not just the approximate value, but the limiting distribution about the approximate value, of the quantity

$$\tilde{M}_n = \sup_{a \leqslant x \leqslant b} \frac{|\hat{f}(x) - f(x)|}{f(x)^{1/2}}. \tag{3.85}$$

For example, for the case $b - a = 1$, $h_n = n^{-1/4}$ and $K(t)$ the rectangular kernel taking the value 1 on $[-1/2, 1/2]$, their result gives, for any fixed z,

$$P\{(\tfrac{1}{2}\log n)^{1/2}(\tilde{M}_n - d_n) < z\} \rightarrow \exp(-2e^{-z}) \text{ as } n \rightarrow \infty \tag{3.86}$$

where the limiting approximate value d_n is given by

$$d_n = (\tfrac{1}{2}\log n)^{1/2} + (2\log n)^{-1/2}(\log 1/4\pi + \log\log n) \tag{3.87}$$

Bickel and Rosenblatt (1973) also suggest that (3.86) could conceivably be used to construct a confidence band for the true density f from the estimated density $\hat{f}$. In practical terms their suggestion is perhaps a little fanciful since, for example, the rate of convergence to infinity of the normalizing factor $(\tfrac{1}{2}\log n)^{1/2}$ is extremely slow; for n equal to a million, the value of the factor is approximately 2.6. The detail of their proof requires the neglect of several terms of order $h^{1/2}$, a quantity that converges to zero very slowly indeed. This discussion is not intended to belittle their remarkable mathematical achievement. The authors themselves point out that 'these asymptotic calculations are to be taken with a grain of salt' and Rosenblatt warns in his survey (1971, p. 1818) against the over-literal interpretation of asymptotic results. Nevertheless, as he goes on to say, asymptotic theorems are useful if treated with care. For example, they can be used as a starting-point for simulation studies or simulation-based procedures, and they may help to give an intuitive feel for the way that a method will behave in practice.

CHAPTER 4

The kernel method for
multivariate data

4.1 Introduction

Up to now, we have concentrated almost exclusively on the estimation of a density underlying a set of univariate observations. Many, if not most, of the important applications of density estimation in fact involve the analysis of multivariate data, and in this chapter the estimation of multivariate densities will be discussed. Most of the attention, as in Chapter 3, will be on the kernel method. Again, this is not intended to imply that the kernel method is the only, or the best, method to use for multivariate data; other methods will be discussed, and compared with the kernel method, in Chapter 5. A particularly important alternative method for multivariate data is the adaptive kernel approach discussed in Section 5.3.

In the multivariate case, the distinction between different possible applications of density estimation becomes more important than for the univariate case discussed in Chapter 3. It is easy to comprehend a contour plot or perspective view of a two-dimensional density function. However, presentational difficulties make it unlikely that density estimates will be useful, directly, for exploratory purposes in more than two dimensions. At a pinch the experienced user with access to sophisticated graphics facilities might be able to inspect and gain profitable insight from a three-dimensional density function. See, for example, Scott and Thompson (1983), who even consider the presentation of four- and five-dimensional densities. On the other hand, if the intention is not to look at the density function but instead to use it as an ingredient in some other statistical technique, then presentational aspects become less important and it may be useful and necessary to estimate densities in higher-dimensional space. Further remarks about specific applications will be made in Chapter 6.

4.2 The kernel method in several dimensions

In this section the kernel method for the multivariate case will be introduced and some comparisons made with multivariate histograms and scatter plots. Throughout the chapter, bold face will be used for points in d-dimensional space. It will be assumed that $\mathbf{X}_1, \dots \mathbf{X}_n$ is the given multivariate data set whose underlying density is to be estimated.

4.2.1 Definition of the multivariate kernel density estimator

The definition of the kernel estimator as a sum of 'bumps' centred at the observations is easily generalized to the multivariate case. The multivariate kernel density estimator with kernel K and window width h is defined by

$$\hat{f}(\mathbf{x}) = \frac{1}{nh^d} \sum_{i=1}^{n} K \left\{ \frac{1}{h}(\mathbf{x} - \mathbf{X}_i) \right\}. \tag{4.1}$$

The kernel function $K(\mathbf{x})$ is now a function, defined for d-dimensional $\mathbf{x}$, satisfying

$$\int_{R^d} K(\mathbf{x}) \, d\mathbf{x} = 1. \tag{4.2}$$

Usually K will be a radially symmetric unimodal probability density function, for example the standard multivariate normal density function

$$K(\mathbf{x}) = (2\pi)^{-d/2} \exp\left(-\tfrac{1}{2}\mathbf{x}^T\mathbf{x}\right). \tag{4.3}$$

Another possible kernel is the multivariate Epanechnikov kernel

$$K_e(\mathbf{x}) = \begin{cases} \tfrac{1}{2}c_d^{-1}(d+2)(1-\mathbf{x}^T\mathbf{x}) & \text{if } \mathbf{x}^T\mathbf{x} < 1 \\ 0 & \text{otherwise} \end{cases} \tag{4.4}$$

where c_d is the volume of the unit d-dimensional sphere: $c_1 = 2, c_2 = \pi$, $c_3 = 4\pi/3$, etc. We shall see in Section 4.4 that useful kernels for the case $d = 2$ are given by

$$K_2(\mathbf{x}) = \begin{cases} 3\pi^{-1}(1-\mathbf{x}^T\mathbf{x})^2 & \text{if } \mathbf{x}^T\mathbf{x} < 1 \\ 0 & \text{otherwise} \end{cases} \tag{4.5}$$

and

$$K_3(\mathbf{x}) = \begin{cases} 4\pi^{-1}(1-\mathbf{x}^T\mathbf{x})^3 & \text{if } \mathbf{x}^T\mathbf{x} < 1 \\ 0 & \text{otherwise.} \end{cases} \tag{4.6}$$

The advantage of these kernels over the Epanechnikov kernel is that the kernels, and hence the resulting density estimates, have higher differentiability properties. In addition, they can be calculated more quickly than the normal kernel (4.3). For further discussion of computational aspects see Section 4.4.

The use of a single smoothing parameter h in (4.1) implies that the version of the kernel placed on each data point is scaled equally in all directions. In certain circumstances, it may be more appropriate to use a vector of smoothing parameters or even a matrix of shrinking coefficients. This will be the case, for example, if the spread of the data points is very much greater in one of the coordinate directions than the others. Just as for many other multivariate statistical procedures, it is probably best to pre-scale the data to avoid extreme differences of spread in the various coordinate directions. If this is done then there will generally be no need to consider more complicated forms of the kernel density estimate than the form (4.1) involving a single smoothing parameter.

An attractive intuitive approach, suggested by Fukunaga (1972, p. 175) is first to 'pre-whiten' the data by linearly transforming them to have unit covariance matrix; next to smooth using a radially

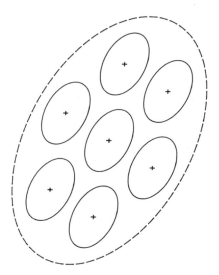

Fig. 4.1 *Shape of kernel and shape of distribution if Fukunaga method is used. After Fukunaga (1972).*

symmetric kernel; and finally to transform back. This is equivalent to using a density estimate of the form

$$\hat{f}(\mathbf{x}) = \frac{(\det S)^{-1/2}}{nh^d} \sum_{i=1}^{n} k\{h^{-2}(\mathbf{x} - \mathbf{X}_i)^T S^{-1}(\mathbf{x} - \mathbf{X}_i)\}, \qquad (4.7)$$

where the function k is given by

$$k(\mathbf{x}^T\mathbf{x}) = K(\mathbf{x})$$

and S is the sample covariance matrix of the data. If, for instance, K is the normal kernel, then $k(u)$ is equal to $(2\pi)^{-d/2} \exp(-\frac{1}{2}u)$. It would perhaps be advisable to use a robust version of the sample covariance for S; see the discussion of 'sphering' data given in Tukey and Tukey (1981, Section 10.2.2). The idea of (4.7), illustrated in Fig. 4.1, is to use a kernel which has the same shape as the data set itself.

4.2.2 Multivariate histograms

The arguments for using density estimates rather than histograms become much stronger in two or more dimensions. The construction of a multivariate histogram requires the specification not only of the size of the bins and the origin of the system of bins, but also the orientation of the bins. In addition there are severe presentational difficulties, even in the two-dimensional case. Figure 4.2, reproduced from Scott (1982), shows a typical bivariate histogram. Partly because

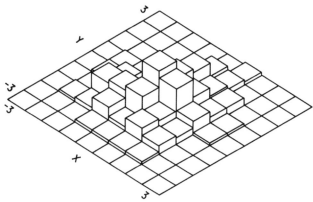

Fig. 4.2 *A typical bivariate histogram. Reproduced from Scott (1982) with the permission of the author.*

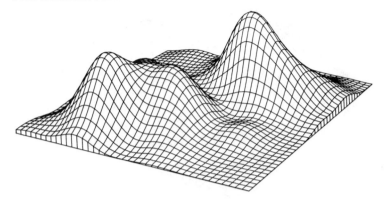

Fig. 4.3 *Density estimate, displayed on the square (±3, ±3), for 100 observations from bivariate normal mixture, window width 1.2.*

of the discontinuous 'block' nature of the bivariate histogram, it is difficult without some experience of looking at diagrams of this type to grasp the structure of the data.

Because they are continuous surfaces, bivariate density estimates constructed using continuous kernels are much easier to comprehend, either as perspective views or as contour plots. For example, Figs 4.3, 4.4 and 4.5, constructed from 100 data points drawn from a bivariate normal mixture, provide a clear picture of the underlying

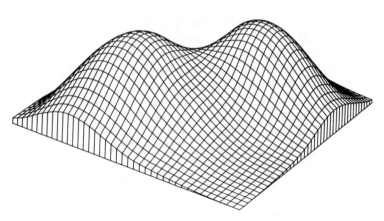

Fig. 4.4 *Density estimate for 100 observations from bivariate normal mixture, window width 2.2.*

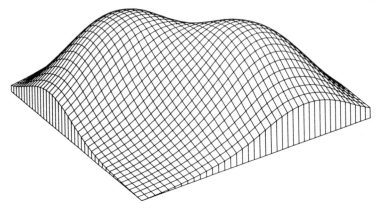

Fig. 4.5 *Density estimate for 100 observations from bivariate normal mixture, window width 2.8.*

distribution. Even the spurious structure of the undersmoothed Fig. 4.3 is quite clear. The difficulty of presenting discontinuous surfaces makes it not at all surprising that although nearly every computer graphics software package has facilities for drawing perspective views of continuous surfaces, very few contain programs for drawing figures like Fig. 4.2. Furthermore, it is of course virtually impossible to draw a meaningful contour diagram of a bivariate histogram.

A further difficulty with histograms is that, if the bin width is chosen small enough to have any chance of catching reasonably local information in any of the coordinate directions, then, even in two dimensions, the total number of bins becomes so large that random error effects are likely to become dominant. This is what has happened in Fig. 4.2. The figure was constructed from 200 observations; the 9×9 grid of histogram bins yields 81 bins altogether, certainly an excessive number for the size of the original data set. A realistic number of bins for the construction of a histogram from 200 observations would be 9 or 16, and the resulting 3×3 or 4×4 bivariate histogram would be very crude. In more than two dimensions the number of bins will rapidly get out of hand. As Tukey and Tukey (1981, p. 223) say about multivariate histograms, '... it is difficult to do well with bins in so few as two dimensions. Clearly, bins are for the birds!'

4.2.3 *Scatter plots*

Of course, an important and obvious step in the examination of a bivariate data set is to look at a scatter plot of the data. However, it is often the case that a density estimate will detect or highlight features that are not at all obvious in the scatter plot. An example is given in Fig. 4.6. These data consist of pairs of readings of plasma lipid concentrations taken on 320 diseased patients in a heart disease study; see Scott *et al.* (1978). These data were very kindly made available to me by David Scott.

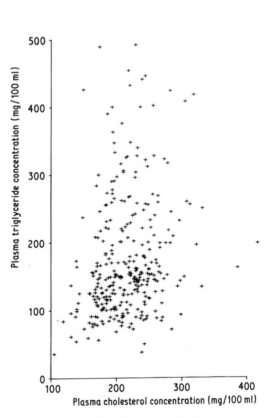

Fig. 4.6 *Scatter plot of plasma lipid data.*

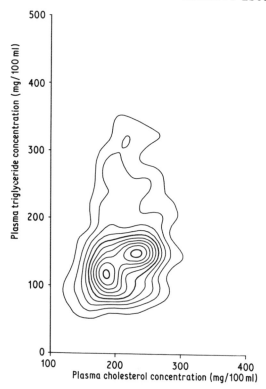

Fig. 4.7 *Kernel estimate for plasma lipid data, window width 40.*

Figure 4.7 is a contour diagram of a density estimate constructed using the kernel K_3 of (4.6) above. The window width is 40, chosen subjectively; the same general structure is visible for a wide range of window widths. The contours are drawn at equal vertical intervals, using the method described at the end of Section 4.4.1. A feature of the data that is very clear from the density estimate is the bimodality in the main part of the distribution. This is hard to see from the scatter plot, even after consideration of the density estimate; an inset showing the dense part of the scatter plot is given as Fig. 4.8. Even if the clustering were clearer from the scatter plots, the density estimate would still have the advantage of providing estimates of the positions of the two modes.

Scott *et al.* (1978) constructed a density estimate similar to Fig. 4.7

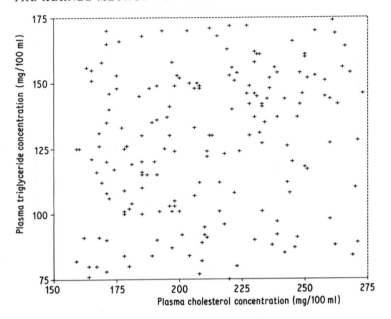

Fig. 4.8 *Inset showing high-density part of Fig. 4.6.*

and used the observed bimodality to divide the population into two groups, by drawing a line perpendicular to the join of the two modes, intersecting this join at the point of smallest estimated density. An interesting clinical difference was found between the two groups.

If the data set under consideration is very large, then it may be expensive in time and ink to produce a scatter plot, and the resulting dense picture may be hard to interpret. The use of automatic data collection methods has increased enormously the number of very large data sets demanding attention from statisticians. An example of a density estimate constructed from a bivariate data set of about 15 000 points is given in Fig. 4.9, reproduced from Silverman (1981a). This data set consists of pairs of observations of the height of the steel surface discussed in Section 1.2, taken at ordered pairs of points $(\mathbf{x}, \mathbf{x} + \mathbf{u})$ separated by a fixed displacement $\mathbf{u}$. The data are scaled so that their mean is at $\mathbf{0}$. It is interesting from the point of view of model-building to note that the mode does not coincide with the mean. In addition, the symmetry of the density about the leading diagonal is important, because it suggests that the statistical properties of the

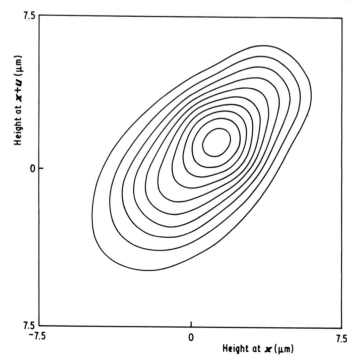

Fig. 4.9 *Kernel estimate for steel surface data, window width 3. Reproduced from Silverman (1981a) with the permission of John Wiley & Sons Ltd.*

surface are apparently invariant under rotation through 180°. For further discussion of this data set see Silverman (1981a) and Adler and Firman (1981).

Scott and Thompson (1983) give a more philosophical argument for using density estimates: '... we take some issue with those who would like to treat the scattergram *sui generis*.... The density function is the fixed entity to which the scattergram merely points.' This is to say that density estimation, however imperfect, is an attempt to discern features of the model underlying the data, not just a convenient presentational device.

4.3 Choosing the kernel and the window width

Very much of the discussion in Chapter 3 concerning the choice of kernel and of window width can be extended, with appropriate modifications, to cover the multivariate case. The technical details of the various changes necessary are reviewed in this section. For fuller discussion the reader is referred back to the relevant sections of Chapter 3.

4.3.1 Sampling properties

As in Section 3.3, approximate expressions can be derived for the bias and the variance of the estimators, and these can be used to give some guidance concerning the appropriate choice of kernel and window width. For rigorous theoretical results underlying this discussion, see Cacoullos (1966) and Epanechnikov (1969). Assume that the kernel K is a radially symmetric probability density function and that the unknown density f has bounded and continuous second derivatives.

Define constants α and β by

$$\alpha = \int t_1^2 K(\mathbf{t})\, d\mathbf{t}$$

and

$$\beta = \int K(\mathbf{t})^2\, d\mathbf{t}. \tag{4.8}$$

Essentially the same manipulations as before, using the multi-dimensional form of Taylor's theorem, yield the approximations

$$\text{bias}_h(\mathbf{x}) \approx \tfrac{1}{2} h^2 \alpha \nabla^2 f(\mathbf{x}) \tag{4.9}$$

and

$$\text{var}\, \hat{f}(\mathbf{x}) \approx n^{-1} h^{-d} \beta f(\mathbf{x}). \tag{4.10}$$

Combining these as in Section 3.3.2 gives the approximate mean integrated square error

$$\tfrac{1}{4} h^4 \alpha^2 \int \{\nabla^2 f(\mathbf{x})\}^2\, d\mathbf{x} + n^{-1} h^{-d} \beta. \tag{4.11}$$

Hence the approximately optimal window width, in the sense of minimizing mean integrated square error, is given by

$$h_{\text{opt}}^{d+4} = d\beta \alpha^{-2} \left\{ \int (\nabla^2 f)^2 \right\}^{-1} n^{-1}. \tag{4.12}$$

The conclusions that can be drawn from this discussion parallel very closely those of Section 3.3 for the univariate case. The approximately optimal window width of (4.12) converges to zero as n increases, but does so extremely slowly, at the rate $n^{-1/(d+4)}$. Furthermore, the appropriate value of h depends on the unknown density being estimated.

The value h_{opt} can be substituted back into (4.11) to give the approximate minimum possible mean integrated square error, and hence, as in Section 3.3.2, to guide the choice of kernel. The Epanechnikov kernel (4.4) is optimum among non-negative kernels in the sense of minimizing the smallest mean integrated square error achievable; however, just as before, the other kernels defined in Section 4.2.1 can achieve very similar mean integrated square errors. Again, it is appropriate to base the choice of kernel on other considerations, notably the remarks made in Section 4.4 concerning computational aspects.

4.3.2 Choice of window width for a standard distribution

The first step beyond choosing the smoothing parameter entirely subjectively is to use the formula (4.12) to provide the appropriate value of the window width when f is a standard density, such as the multivariate normal. If ϕ is the unit d-variate normal density, then it can be shown that

$$\int (\nabla^2 \phi)^2 = (2\sqrt{\pi})^{-d}(\tfrac{1}{2}d + \tfrac{1}{4}d^2). \tag{4.13}$$

The value given by (4.13) can then be substituted back in (4.12) to give the optimal window width for the smoothing of normally distributed data with unit variance. The window width will be given by

$$h_{opt} = A(K)n^{-1/(d+4)} \tag{4.14}$$

where the constant

$$A(K) = \left[d\beta\alpha^{-2} \left\{ \int (\nabla^2 \phi)^2 \right\}^{-1} \right]^{1/(d+4)} \tag{4.15}$$

depends on the kernel and is tabulated in Table 4.1.

Consider, now, the smoothing of a general data set with covariance matrix S, possibly robustly estimated. If the Fukunaga estimate (4.7) is being used, then the h_{opt} of (4.14) gives a directly appropriate value for the smoothing parameter. On the other hand, if the kernel is radially symmetric and the data are not transformed, the procedure indicated

Table 4.1 *The constant $A(K)$, as defined in equation (4.15), for various kernels* *K. Numerical values are given correct to 2 decimal places*

Kernel	Dimensionality	$A(K)$
Multivariate normal K	2	1
as in equation (4.3)	d	$\{4/(d+2)\}^{1/(d+4)}$
K_e as in equation (4.4)	2	2.40
	3	2.49
	d	$\{8c_d^{-1}(d+4)(2\sqrt{\pi})^d\}^{1/(d+4)}$
K_2 as in equation (4.5)	2	2.78
K_3 as in equation (4.6)	2	3.12

would be to choose a single scale parameter σ for the data and to use the value σh_{opt} for the window width. A possible choice for σ might be

$$\sigma^2 = d^{-1} \sum_i s_{ii},$$

so that σ^2 is the average marginal variance.

The cautionary remarks made in Section 3.4.2 about the use of the window width $1.06\sigma n^{-1/5}$ for the univariate case apply more strongly for the corresponding multivariate value given by (4.14), and in particular it will often be appropriate to use a slightly smaller value. Nevertheless, the method described above does give a quick and easy choice of at least an initial value for the window width.

4.3.3 *More sophisticated ways of choosing the window width*

Both least-squares cross-validation and likelihood cross-validation carry over to the multivariate case with no essential modification at all. The least-squares cross-validation score $M_1(h)$ of (3.39) becomes

$$M_1(h) = n^{-2}h^{-d} \sum_i \sum_j K^*\{h^{-1}(\mathbf{X}_i - \mathbf{X}_j)\} + 2n^{-1}h^{-d}K(\mathbf{0}). \quad (4.16)$$

It is interesting to note that the computational effort required to calculate $M_1(h)$ from (4.16) depends on the dimensionality d only in the time taken to calculate the squared distances $(\mathbf{X}_i - \mathbf{X}_j)^{\mathrm{T}}(\mathbf{X}_i - \mathbf{X}_j)$. The theory of Stone (1984) applies equally to the multivariate case.

The difficulty caused in likelihood cross-validation by outlying observations becomes more serious when multivariate data are considered, both because it becomes more difficult to detect outliers and because there is, loosely speaking, more space in which outliers can occur. Advocates of the method (see, for example, Habbema, Hermans and van der Broek, 1974) have reported quite good results using kernels of unbounded support, but these kernels are likely to require high computation times. For this reason, and because of the likely dependence on tail properties of the kernel discussed in Section 3.4.4, least-squares cross-validation is probably to be preferred.

The test graph method of Section 3.4.5 can, in principle, be extended, though it is not likely to be practicable in more than two dimensions. In two dimensions, the method requires the consideration of 'test surfaces' giving plots of $\nabla^2 \hat{f}$. Producing these is expensive computationally; nevertheless, an example given in Silverman (1978a, Section 4) suggests that the method may be of some value, at least as a further check on a window width chosen by another method. Certainly the test surface is more sensitive to small changes in h than is the density estimate itself.

Finally, the iterative approach of Scott, Tapia and Thompson (1977) described in Section 3.4.6 can, in principle, be used in the multivariate case though, as Scott (1982) points out, the computational burden is likely to be heavier. In addition, estimating $\int \nabla^2 f$ by $\int \nabla^2 \hat{f}$ seems somewhat hazardous, unless very large samples are available.

4.4 Computational considerations

Even when using a high-speed computer it is important to take some care in calculating multivariate density estimates. Especially when the sample size is large, use of an inappropriate algorithm can lead to excessively long calculation times.

One important factor in reducing the computer time is the choice of a kernel that can be calculated very quickly. If the sample size is 1000 and the formula (4.1) is used directly to find the estimate at each point of a grid of 900 points, then, of course, nearly a million evaluations of the kernel function will be required. In these circumstances, every arithmetic operation that can be saved in the calculation of the kernel will make a noticeable difference. In particular, it would be foolish to use a kernel like the normal density (4.3), which requires a call to the exponential function at each evaluation. It is obvious, but often overlooked, that it is more efficient to take any common multiplicative factors out of the inner summation in which the kernel is evaluated, and to rescale the estimate after all the data points have been considered. Thus, for instance, if the Epanechnikov kernel (4.4) is being used in the case $d = 2$, it is best to find the function

$$\sum_i \left\{ 1 - \left(\frac{\mathbf{X}_i - \mathbf{t}}{h}\right)^T \left(\frac{\mathbf{X}_i - \mathbf{t}}{h}\right) \right\}$$

and then to multiply by the precalculated constant $2/(\pi n h^2)$ to find the estimate $\hat{f}(\mathbf{t})$.

4.4.1 Regular grids : contour diagrams and perspective views

Nearly all package programs for finding contour diagrams and perspective views of functions require, as input, an array giving the value of the function at each point of a regular grid. Suppose we are working in the two-dimensional case, and that the (j, k)th point $\mathbf{t}_{jk}$ of the grid has coordinates $(x_{l_o} + j\delta, y_{l_o} + k\delta)$ for some origin (x_{l_o}, y_{l_o}). Assume that the kernel satisfies $K(\mathbf{x}) = 0$ if $\mathbf{x}^T\mathbf{x} \geqslant 1$; for example the kernels (4.4) and (4.5) have this property.

A data point at $\mathbf{x}_1 = (x, y)$ can only contribute to the value of the density estimate at points $\mathbf{t}_{jk}$ satisfying

$$\begin{align} \delta^{-1}(x - x_{l_o}) - h\delta^{-1} &< j < \delta^{-1}(x - x_{l_o}) + h\delta^{-1} \\ \delta^{-1}(y - y_{l_o}) - h\delta^{-1} &< k < \delta^{-1}(y - y_{l_o}) + k\delta^{-1} \end{align} \tag{4.17}$$

since otherwise the distance between $\mathbf{t}_{jk}$ and $\mathbf{x}$ will be greater than h. It is best to consider each data point in turn and find the range of values of j and k given by (4.17). The contribution of the data point to the density estimate at *all* these grid points is then added to the array of function values, before going on to consider the next data point. This strategy reduces considerably the number of evaluations of the kernel, because a very large number of pairs $(\mathbf{t}_{jk}, \mathbf{X}_i)$ are never considered at all.

It is obviously worth using as coarse a grid as possible to reduce the number of evaluations of $\hat{f}(\mathbf{t})$ and also to economize on the post-processing involved in determining the contours or perspective view from the grid of function values. In most packages this final step involves some form of linear interpolation between the grid points. A typical method is to divide each grid square along the diagonals into four triangles, to estimate the value at the centre as the average of the four corner values, and then to fit a linear surface on each of the four triangles. The effect of this kind of procedure is to approximate the function f by a piecewise linear function f_{approx}, which depends locally on f.

If the grid interval is δ, it can be shown by manipulations involving Taylor's theorem that the maximum difference between f and f_{approx} will be $O(\delta^2)$ as δ tends to zero, provided f is differentiable everywhere and has bounded second derivatives. In contrast to most applications of contouring, the smoothness properties of the density estimate $\hat{f}$, because they are determined by those of the kernel, are entirely under the user's control. For example, the kernel K_2 of (4.5) has the required smoothness properties, while the Epanechnikov kernel (4.4) does not. The practical implication of this discussion is that, provided a kernel such as K_2 is used, a reasonably coarse grid can be used for contouring the kernel estimate without introducing appreciable errors.

Still greater economies are possible if a contouring method that makes use of gradient information is available. A special feature of density estimation is that the gradients of the estimate are given explicitly, by the formula

$$\nabla \hat{f}(\mathbf{t}) = n^{-1}h^{-d-1}\sum_i \nabla K\{h^{-1}(\mathbf{t} - \mathbf{X}_i)\}. \tag{4.18}$$

This is not the case for most surface plotting problems, where gradients are often unavailable or else expensive to obtain. If a kernel like K_3 is used, then it takes little additional effort to use the formula (4.18) to find the gradients of the estimate at each grid point at the same time as calculating the estimate itself. The values *and* gradients are then fed to the contouring package.

Silverman (1981a) uses a contouring method of Sibson and Thomson (1981) for this purpose; see also Thomson (1984). This method matches the given values and gradients to those of a smooth piecewise quadratic surface. Provided the kernel has the high-order smoothness properties of the kernel K_3 of (4.6), the approximation errors are reduced to $O(\delta^3)$ and so interpolation even from quite large grid squares is still acceptably accurate.

All the bivariate density estimates depicted in this chapter were obtained from coarse grids using this technique. Figure 4.3 was obtained from an 11×11 grid of values and gradients, while the 6×6 grids used for the smoother Figs 4.4 and 4.5 gave results which were visually indistinguishable from those obtained using finer grids. Using a 6×6 grid for the surfaces of Fig. 4.3 made a noticeable but not substantial difference.

4.4.2 Evaluation at irregular points

In some bivariate applications, and in most applications in three or more dimensions, it will be necessary to find the value of the density estimate at various irregularly or randomly placed points, rather than to plot a picture of the whole function from a grid of values.

Suppose that a kernel K satisfying $K(\mathbf{x}) = 0$ for $\mathbf{x}^T\mathbf{x} \geqslant 1$ is being used. If the sample size n is very large, it may be worth preprocessing the data points by ordering them on the values of one of the coordinates. Write $\mathbf{X}_i = (X_{i1}, \ldots, X_{id})$ for each i and suppose that the sorting has been done on the first coordinate. Given a point $\mathbf{t} = (t_1, \ldots, t_d)$ at which the density estimate is required, let i_{lo} be the first value of the index i for which $X_{i1} > t_1 - h$, and let i_{hi} be the last i for which $X_{i1} < t_1 + h$. The indices i_{lo} and i_{hi} can be found by any suitable rapid search procedure. The density estimate can then be written and calculated as

$$\hat{f}(\mathbf{t}) = n^{-1}h^{-d} \sum_{i=i_{lo}}^{i_{hi}} K\{h^{-1}(\mathbf{t} - \mathbf{X}_i)\} \tag{4.19}$$

since $\mathbf{X}_i$ with i outside the range (i_{lo}, i_{hi}) will not contribute to the density estimate at $\mathbf{t}$.

The efficiency of this procedure is enhanced by a suitable choice of the coordinate direction on which to sort the data. It is best to choose the direction in which the spread of the data is greatest. The originators of the procedure, Friedman, Baskett and Shustek (1975), give further details and discussion, and even suggest a method where the chosen coordinate direction is allowed to depend on the point $\mathbf{t}$.

4.5 Difficulties in high-dimensional spaces

It is well known, but often ignored, in multivariate analysis generally that intuition gained through everyday practical experience in one, two or three dimensions may be misleading when considering data in higher-dimensional spaces. In the context of density estimation, the peculiarities of high-dimensional spaces manifest themselves in several ways. Some awareness of the problems should discourage users of the techniques from attempting to apply them in inappropriate circumstances. Although the actual definition of probability density does not change as the dimensionality changes, there are subtle differences that are likely to make density estimation difficult. An illustration of the failure of one-dimensional intuition in high-dimensional spaces is given in the next section.

4.5.1 *The importance of the tails in high dimensions*

Given almost any standard density f, consider the effect of resetting $f(\mathbf{x})$ to zero at all points where $f(\mathbf{x})$ is less than 0.01 sup f, and rescaling the resulting function to be a probability density. In one dimension this will make very little difference for almost all densities except those with very sharp peaks, since very few data points will occur in regions where f is very small. Thus samples of moderate size drawn from f will look similar to those drawn from its truncated version; in density estimation it will not be crucial whether the tails are estimated particularly accurately.

In ten dimensions, on the other hand, the picture is very different. Concentrate attention on the standard normal distribution, ostensibly one of the best behaved and most short-tailed densities. Even for the normal over half the observations will, on average, fall at points where the density $f(\mathbf{x})$ is less than one-hundredth of its maximum value. (This is a consequence of the fact that, if $\mathbf{X}$ is standard ten-variate normal,

$$f(\mathbf{X})/f(\mathbf{0}) = \exp\left(-\tfrac{1}{2}\mathbf{X}^{\mathrm{T}}\mathbf{X}\right)$$
$$\sim \exp\left(-\tfrac{1}{2}\chi_{10}^2\right);$$

since the median of the χ_{10}^2 distribution is 9.34, the median of $f(\mathbf{X})/f(\mathbf{0})$ is $\exp\left(-9.34/2\right) = 0.0094$.) It follows that the apparently innocuous truncation described in the last paragraph will have an enormous effect. Put another way, this example shows that, in contrast to the low-dimensional case, regions of relatively very low density can still be extremely important parts of the distribution.

Conversely, and perhaps paradoxically, apparently large regions of high density may be completely devoid of observations in a sample of moderate size. Scott and Thompson (1983) call this the *empty space phenomenon*. Sticking with the ten-dimensional normal example for the moment, a similar argument to that given above shows that 99% of the mass of the distribution is at points whose distance from the origin is greater than 1.6. This is in stark contrast to the one-dimensional case, where nearly 90% of the distribution lies between ± 1.6, and shows both that it is likely to be difficult to estimate the density except from enormous samples, and that the density itself may give a superficially false impression of the likely behaviour of sample data sets. Even if the tail of the density is eliminated altogether by considering the uniform distribution over a box, similar 'empty space' behaviour is observed. If B is the ten-dimensional hypercube $\{\mathbf{x} : |x_i| \leqslant 1 \text{ all } i\}$, then a point distributed uniformly over B will have probability only 0.01 of falling in the hypercube centred at $\mathbf{0}$ with sides of half-length 0.63.

4.5.2 *Required sample sizes for given accuracy*

In order to gain some intuition about when it *is* appropriate to estimate high-dimensional densities, let us consider a very special case. To avoid difficulties about which measure of global accuracy to use, we shall concentrate on the estimation of a density at a single point. The particular case considered has been chosen because it leads to tractable calculations. Suppose, then, that the true density f is unit multivariate normal, and that the kernel is normal. Suppose that it is of interest to estimate f at the point $\mathbf{0}$, and that the window width h has been chosen to minimize the mean square error at this point. A not unreasonable aim would be to ensure that the relative mean square error $E\{\hat{f}(\mathbf{0}) - f(\mathbf{0})\}^2 / f(\mathbf{0})^2$ is fairly small, say less than 0.1. Table 4.2 gives, as a function of dimension, the sample size required to achieve this object. The table was constructed by numerical minimization over h of the exact expression for the mean square error at $\mathbf{0}$, to give the minimum possible mean square error for each sample size. The entries in the table are accurate to about 3 significant figures.

Striking conclusions can be drawn from the table. To get the kind of accuracy available in one and two dimensions from very small samples, and in three and four dimensions from moderate samples, nearly a million observations are needed in ten dimensions. Furthermore, the results obtained in this case are likely to be optimistic since the point $\mathbf{0}$ is not in the tail of the distribution, and the normal is a

Table 4.2 *Sample size required (accurate to about 3 significant figures) to ensure that the relative mean square error at zero is less than 0.1, when estimating a standard multivariate normal density using a normal kernel and the window width that minimizes the mean square error at zero*

Dimensionality	Required sample size
1	4
2	19
3	67
4	223
5	768
6	2 790
7	10 700
8	43 700
9	187 000
10	842 000

smooth unimodal density. It is perhaps surprising just how quickly the required sample size increases with dimension.

Epanechnikov (1969) includes a table giving broadly similar conclusions for the mean integrated square error in up to five dimensions. Unfortunately, some calculations suggest that the values in Epanechnikov's table are incorrect. In all dimensions up to 10, the sample sizes required to give a value of 0.1 for the relative mean integrated square error $E\int(\hat{f} - f)^2/\int f^2$ are approximately 1.7 times those given in Table 4.2. If a measure of global fit more sensitive to tail behaviour were used, then the sample sizes required would, no doubt, be larger still.

CHAPTER 5

Three important methods

5.1 Introduction

In this chapter some methods other than the kernel method will be discussed in detail. Probably the only practical drawback of the kernel method of density estimation is its inability to deal satisfactorily with the tails of distributions without oversmoothing the main part of the density. This was a problem discussed briefly in Section 2.4 and illustrated in Fig. 2.9 in the context of the suicide data. Two possible adaptive approaches, the *nearest neighbour* and *variable kernel* methods, were mentioned in Sections 2.5 and 2.6, though neither of these methods was entirely satisfactory. In Section 5.2 the nearest neighbour method will be considered, and in Section 5.3 an extension of the variable kernel method, called the *adaptive kernel* method, will be introduced and discussed. It will be seen that the adaptive kernel method has certain potential practical advantages over both the kernel and the nearest neighbour methods as a method for smoothing long-tailed distributions. It would be extremely rash to make strong recommendations about which method of density estimation is 'best' since the appropriate choice of course depends on the context. The author's personal view is that the kernel method is a good choice for many practical purposes; it is simple and intuitively appealing, and its mathematical properties are quite well understood. If undersmoothing in the tails is likely to cause difficulties, then the adaptive kernel approach is well worth considering.

A philosophical difficulty with the kernel method, and with most of the other methods for density estimation, is the rather *ad hoc* way in which the estimates are defined. An attempt to place density estimation on a somewhat firmer philosophical footing is described in Section 5.4, where the *maximum penalized likelihood* method is discussed. The practical appeal of penalized likelihood in the specific context of density estimation is probably somewhat less than that of the other methods discussed in this chapter, particularly in the

95

multivariate case. However, the intellectual merits of the method will, it is hoped, provide an impetus for its further practical use and development as computer time becomes increasingly cheap and available numerical maximization algorithms progressively more sophisticated.

Apart from the kernel method and the three methods discussed in this chapter, there are of course many other methods available for density estimation, such as the orthogonal series method mentioned in Section 2.7. These other methods will not be dealt with in detail in this book, but extensive descriptions of various methods together with lists of references are available, for example, in the survey papers and books mentioned in Section 1.3.

5.2 The nearest neighbour method

The nearest neighbour method is widely used in the fields of pattern recognition and nonparametric discriminant analysis. It should more properly be called the near neighbour method since the density estimate depends on near neighbours rather than nearest neighbours of a particular point. It has also been called the balloon density, or balloonogram; see Tukey and Tukey (1981, Section 11.3.2).

5.2.1 Definition and properties

To define the nearest neighbour estimate in d dimensions, let $r_k(t)$ be the Euclidean distance from t to the kth nearest data point, and let $V_k(t)$ be the (d-dimensional) volume of the d-dimensional sphere of radius $r_k(t)$; thus $V_k(t) = c_d r_k(t)^d$, where c_d is the volume of the unit sphere in d dimensions (so that $c_1 = 2$, $c_2 = \pi$, $c_3 = 4\pi/3$, etc.). The nearest neighbour density estimate is then defined by

$$\hat{f}(t) = \frac{k/n}{V_k(t)} = \frac{k/n}{c_d r_k(t)^d}. \tag{5.1}$$

The discussion in Section 2.5 about the motivation for this estimate extends easily to the multivariate case. Of a sample of size n, one would expect about $nf(t)V_k(t)$ observations to fall in the sphere of radius $r_k(t)$ centred at t. Setting this number equal to the number, k, actually observed in this sphere gives the estimate (5.1).

Another related argument given in Section 2.5 reinforces the definition (5.1) and establishes an important connection with kernel estimation.

Consider the kernel estimate based on the kernel

$$K(\mathbf{x}) = \begin{cases} c_d^{-1} & \text{if } |\mathbf{x}| \leq 1 \\ 0 & \text{otherwise}. \end{cases} \tag{5.2}$$

Then (5.1) is precisely the kernel estimate evaluated at $\mathbf{t}$ with window width $r_k(\mathbf{t})$. The generalized kth nearest neighbour estimate (2.4) pursues this connection further by allowing use of more general kernel functions. In the multivariate case, the generalized nearest neighbour estimate is defined, for a general kernel K, by

$$\hat{f}(\mathbf{t}) = n^{-1} r_k(\mathbf{t})^{-d} \sum_{i=1}^{n} K\{r_k(\mathbf{t})^{-1}(\mathbf{t} - \mathbf{X}_i)\}. \tag{5.3}$$

Putting K equal to the function (5.2) gives the estimate (5.1) as a special case.

The connection between kernel and nearest neighbour methods demonstrates that in a practical sense, there is nothing to choose between kernel and nearest neighbour estimates when estimating a density at a single point. Every value k of the smoothing parameter in the nearest neighbour estimate will give an estimate identical to that obtained with a certain value of h in the kernel estimate. It is only when estimating the density at several points or when constructing an estimate of the entire density function that the two methods will give different results.

Unfortunately, as was pointed out in Section 2.5 and illustrated in Figure 2.10, the overall estimates obtained by the nearest neighbour method are not very satisfactory. They are prone to local noise and also have very heavy tails and infinite integral. Using a generalized nearest neighbour estimate (2.4) with a smooth kernel does not help matters very much, because the local irregularities are caused by the dependence of the estimator on the non-differentiable function $r_k(\mathbf{t})$.

As $|\mathbf{t}|$ becomes large, $r_k(\mathbf{t})$ increases as $|\mathbf{t}|$ and so the tails of the ordinary nearest neighbour estimate will decrease as $|\mathbf{t}|^{-d}$, regardless of the tail behaviour of the data. The generalized nearest neighbour estimate may have more reasonable tail behaviour if the kernel function K is smooth and radially symmetric and is zero on and outside the unit sphere. In this case the tails of $\hat{f}(\mathbf{t})$ will decrease more rapidly as $|\mathbf{t}|$ becomes large. However, their exact behaviour will depend in a fairly subtle way on the behaviour of $K(\mathbf{t})$ for $|\mathbf{t}|$ near one and there is no guarantee that $\hat{f}(\mathbf{t})$ will integrate to one; usually it will not. Details are left to the more mathematically inclined reader to investigate.

Some theoretical investigation of the tail properties of generalized nearest neighbour estimates was carried out by Mack and Rosenblatt (1979). Under the assumptions of the last paragraph on K, and assuming that f is continuously twice differentiable, they showed that

$$\text{bias}\{\hat{f}(\mathbf{t})\} \approx \tfrac{1}{2}c_d^{-2/d}\int x_1^2 K(\mathbf{x})\,d\mathbf{x}\left(\frac{k}{n}\right)^{2/d}\frac{\nabla^2 f(\mathbf{t})}{f(\mathbf{t})^{2/d}} \tag{5.4}$$

and

$$\text{var}\{\hat{f}(\mathbf{t})\} \approx c_d\int K(\mathbf{x})^2\,d\mathbf{x}\frac{f(\mathbf{t})^2}{k} \tag{5.5}$$

where c_d is the volume of the unit d-dimensional sphere. Close examination of these expressions for various standard densities f demonstrates that although the estimates are indeed successful in reducing the variance in the tails, they do so at the expense of introducing excessive bias. Rosenblatt (1979) studied measures of the global behaviour of generalized nearest neighbour estimates, such as mean integrated squared error, and found that these are dominated by the tails because of the large contribution made by the bias.

The conclusion of the discussion of this section and of Section 2.5 is that, although the kernel method undersmooths the tails, the nearest neighbour method perhaps overcompensates for this difficulty by smoothing the tails too much. In addition it provides estimates which are not everywhere differentiable but which are subject to local irregularities. The adaptive kernel method discussed in Section 5.3 can, with suitable choice of parameters, overcome these difficulties while still being adaptive to the local density.

5.2.2 Choice of smoothing parameter

The question of how to choose the smoothing parameter has not been investigated at all closely for any method other than the kernel method. In the case of the generalized nearest neighbour method, the expressions (5.4) and (5.5) can be used to see how the ideal smoothing parameter k would behave for a known density f. This provides results analogous to those obtained in Section 3.2.2 for the kernel method.

Comparison of (5.4) and (5.5) with the corresponding results (4.9) and (4.10) for the kernel method shows that the approximate bias and variance for the two methods, at the point **t**, are exactly the same if we choose

$$\frac{k}{n} = c_d h^d f(\mathbf{t}). \tag{5.6}$$

Of course the adaptive nature of the nearest neighbour method makes it inevitable that the correspondence (5.6) should depend on **t**. Nevertheless, it is clear that $(k/n)^{1/d}$ plays the same role as the bandwidth h, and an argument similar to that leading to (4.12) shows that the ideal k for any particular **t** would be proportional to $n^{4/(d+4)}$, the constant of proportionality depending on **t**.

Some of the difficulties alluded to by Mack and Rosenblatt (1979) become clear if an attempt is made to obtain a 'standard' choice for k based on mean integrated squared error for the normal distribution, as was done in Section 3.4.2 for the kernel method. In the one-dimensional case, the approximate integrated square bias, obtained by squaring (5.4) and integrating, involves

$$\int_{-\infty}^{\infty} f''(t)^2 f(t)^{-4} \, dt,$$

which is easily seen to be infinite.

Methods like cross-validation could be used for choosing the smoothing parameter in the nearest neighbour method. However, the behaviour of the estimates is such that they are best used in applications where the choice of smoothing parameter is not very critical anyway. It would probably be best to choose k within the context of the particular application by trying several different values and choosing the one which gives the most satisfactory results.

5.2.3 Computational hints

One of the reasons for the popularity of the nearest neighbour method in areas related to computer science is the amount of work that has been done on fast algorithms for finding near neighbours. An example of such a technique is the preprocessing suggested by Friedman, Baskett and Shustek (1975) and described in Section 4.4.2 above. A more sophisticated technique is described by Friedman, Bentley and

Finkel (1977). In this method, the data points are successively partitioned and arranged in a tree structure. At each node of the tree, the data points below that node are split into two subsets along one of the coordinate directions; the choice of coordinate direction may vary from one node to another, and the authors suggest choosing the direction in which the points are most spread out. By using an appropriate data structure, the k nearest neighbours in the data set of any particular test point can be found extremely rapidly because large parts of the data set are excluded from consideration almost at once. Full details of the method are given in the published reference. The use of a sophisticated technique of this kind is probably only worth contemplating if a fairly large number of large data sets are to be processed. Otherwise a more naive approach will usually be good enough.

It should be stressed that fast algorithms for finding near neighbours can also be used for finding kernel estimates, since (provided a kernel of bounded support is used) these also depend on near neighbours of the point under consideration. The only difference is that, instead of finding the k nearest neighbours of a point, it is necessary to find all the data points within a given distance of that point; the algorithm of Friedman, Bentley and Finkel (1977) is easily adapted to do this. The algorithm can also be used to find the adaptive kernel estimates defined in the next section.

5.3 Adaptive kernel estimates

The general aim of this section is to develop a class of estimates that combines features of the kernel and nearest neighbour approaches. The basic idea is to construct a kernel estimate consisting of 'bumps' or kernels placed at the observed data points, but to allow the window width of the kernels to vary from one point to another. This procedure is based on the common-sense notion that a natural way to deal with long-tailed densities is to use a broader kernel in regions of low density. Thus an observation in the tail would have its mass smudged out over a wider range than one in the main part of the distribution.

The discussion of Section 4.5 shows that the tail of the distribution is particularly important in the multivariate case. Although it is futile to expect very good estimates to be obtained in high-dimensional spaces, there are many important problems involving data in, say, two, three or four dimensions; analogous arguments to those of

Section 4.5.1 show that the regions of relatively low density will be of far more importance than one-dimensional intuition could lead us to expect.

5.3.1 *Definition and general properties*

An obvious practical problem is deciding in the first place whether or not an observation *is* in a region of low density. The adaptive kernel approach copes with this problem by means of a two-stage procedure. An initial estimate is used to get a rough idea of the density; this estimate yields a pattern of bandwidths corresponding to the various observations and these bandwidths are used to construct the adaptive estimator itself. The general strategy used will be as follows (assume that the data points lie in d-dimensional space):

Step 1 Find a *pilot estimate* $\tilde{f}(\mathbf{t})$ that satisfies $\tilde{f}(\mathbf{X}_i) > 0$ for all i.

Step 2 Define *local bandwidth factors* λ_i by

$$\lambda_i = \{\tilde{f}(\mathbf{X}_i)/g\}^{-\alpha} \tag{5.7}$$

where g is the geometric mean of the $\tilde{f}(\mathbf{X}_i)$:

$$\log g = n^{-1} \sum \log \tilde{f}(\mathbf{X}_i)$$

and α is the *sensitivity parameter*, a number satisfying $0 \leqslant \alpha \leqslant 1$.

Step 3 Define the *adaptive kernel estimate* $\hat{f}(\mathbf{t})$ by

$$\hat{f}(\mathbf{t}) = n^{-1} \sum_{i=1}^{n} h^{-d} \lambda_i^{-d} K\{h^{-1} \lambda_i^{-1}(\mathbf{t} - \mathbf{X}_i)\} \tag{5.8}$$

where K is the kernel function and h is the bandwidth. As in the ordinary kernel method, K is a symmetric function integrating to unity.

Various remarks about this strategy can be made. The first step, the construction of the pilot estimate, requires the use of another density estimation method, such as the kernel or nearest neighbour method. The general view in the literature (see Breiman, Meisel and Purcell, 1977; Abramson, 1982), confirmed by the present author's experience, is that the method is insensitive to the fine detail of the pilot estimate, and therefore any convenient estimate can be used. A natural pilot estimate would be a fixed kernel estimate with bandwidth chosen by reference to a standard distribution, as in

Sections 3.4.2 and 4.3.2. There is no need for the pilot estimate to have any particular smoothness properties; therefore a natural kernel to use in the multivariate case is the Epanechnikov kernel. It would certainly be unwarranted to use any time-consuming method like cross-validation in the construction of the pilot estimate.

Allowing the local bandwidth factors to depend on a power of the pilot density gives flexibility in the design of the method. The larger the power α, the more sensitive the method will be to variations in the pilot density, and the more difference there will be between bandwidths used in different parts of the sample. The value $\alpha = 0$ will reduce the method back to the fixed width kernel approach, since all the λ_i will then be equal to one. Further remarks about the choice of α will be made in Section 5.3.3. We shall see that there are good reasons for setting $\alpha = \frac{1}{2}$.

Including the factor g^α in (5.7) is not strictly speaking necessary, since it could be absorbed into the bandwidth h used in the final stage. However, it has the advantage of freeing the bandwidth factors from the scale of the data and imposing the constraint that the geometric mean of the λ_i is equal to one. In the final stage, the width of the kernel placed at $\mathbf{X}_i$ is equal to $\lambda_i h$. If the λ_i are constructed as in (5.7) then the bandwidth h controls the amount of smoothing applied generally to the data. It might be appropriate to construct the pilot estimate to be sensitive to effects on the same sort of scale as the final estimate; this would suggest, for example, using as the pilot estimate a fixed kernel estimate of bandwidth h, the same as for the final estimate. This procedure has been found by the author to give good results, but it has the disadvantage of requiring the recalculation of the pilot estimate for each new value of the bandwidth h.

The definition (5.8) of the adaptive kernel estimate ensures that, provided K is non-negative, the estimate will be a bona fide probability density and will not suffer from excessively heavy tails in the way that the nearest neighbour method does. Furthermore, the estimate will inherit all the differentiability properties of the kernel; if derivatives of the estimate are required, for example, for purposes of interpolation from a grid as in Section 4.4.1, they are available by differentiating (5.8) with respect to $\mathbf{t}$.

5.3.2 The method of Breiman, Meisel and Purcell

These authors, in their 1977 paper, suggested and investigated an estimate which is a special case of the adaptive kernel estimate and which was defined and discussed briefly in Section 2.6 above. In the terminology of the adaptive kernel framework, they used as a pilot estimate the nearest neighbour estimate with a fairly large value of the smoothing parameter k, and then set the sensitivity parameter α equal to $1/d$, where d is the dimensionality of the space in which the density is being estimated. We shall see below that this may not always be a very good choice of α, but nevertheless their general conclusions are very interesting. They found in a simulation study that the quality of their variable kernel estimate was 'surprisingly insensitive' to the choice of smoothing parameter in the pilot estimate $\tilde{f}$. Judged by the sum of squared errors at the data points, the variable kernel method was capable of giving results almost twice as good as those of the fixed width kernel method, if the bandwidth was chosen optimally in both cases. As might be expected, the advantages of the variable kernel method were found to be greater in the case where the density itself is more variable, but even in the case of normal f the improvement was dramatic.

5.3.3 Choice of the sensitivity parameter

The choice of sensitivity parameter $\alpha = 1/d$ used by Breiman, Meisel and Purcell (1977) will ensure that the number of observations 'caught' by the scaled kernel will be approximately the same in all parts of the density. To see this, notice that, provided h is small, the expected number of observations in a ball of radius $hf(\mathbf{t})^{-1/d}$ centred at $\mathbf{t}$ is approximately equal to $f(\mathbf{t}) \times$ (volume of the ball) $= c_d h^d$, where c_d is the volume of the unit d-dimensional sphere. Though a choice of this kind, reminiscent of the construction of the nearest neighbour estimate, has some intuitive appeal, it is worth considering the choice of α somewhat more carefully.

Concentrate on the univariate case for the moment. Consider the fixed-width kernel estimate and suppose it is of interest to get a good estimate at the single point x. Assume K is symmetric and non-negative. Simple calculus based on the approximations (3.16) and (3.18) shows that the asymptotic choice of bandwidth to minimize the mean square error at x is given by

$$h(x) = C_1(K)n^{-1/5} |f''(x)|^{-2/5} f(x)^{1/5} \tag{5.9}$$

This is the formula of Parzen (1962, equation 4.15); the quantity $C_1(K)$ is a constant depending only on the kernel. The formula (5.9) depends on both f and on f''. If we presume that the likely values of

f'' are proportional to those of f (i.e. lower values of f'' in the tails) then (5.9) suggests that the ideal local bandwidth will be proportional to $f^{-1/5}$. This is a rather heuristic argument depending on asymptotic calculations and possibly rash assumptions, but it does suggest that a value of the sensitivity parameter α rather less than 1 may be appropriate.

A very interesting argument in favour of setting α equal to $1/2$ is given by Abramson (1982). Suppose that the estimate (5.8) is used with λ_i *equal* to $f(X_i)^{-1/2}$, in other words that errors involved in the pilot estimation can be ignored. (Abramson gives some justification of this assumption.) Consider the bias in $\hat{f}$ at the point t. We will have

$$E\hat{f}(t) = \int h^{-1} f(x)^{1/2} K\{(t-x) \, f(x)^{1/2} h^{-1}\} f(x) \, dx$$

$$= \int f(t+hu)^{3/2} K\{uf(t+hu)^{1/2}\} \, du \qquad (5.10)$$

substituting $u = (x-t)/h$ and using the symmetry of K. In order to obtain an approximation to (5.10) valid for small h, a Taylor series expansion for the integrand can be used. Abramson (1982) demonstrated the remarkable fact that, in contrast to the fixed-width kernel estimate with kernel K, the $O(h^2)$ term in the Taylor expansion is identically zero. Thus using bandwidth factors inversely proportional to the square root of the pilot estimate will give an estimate whose bias is of smaller order than that of the fixed-width kernel estimate. Abramson also shows that the same result holds in the multivariate case, and furthermore that no other dependence of the local bandwidth on $\tilde{f}$ will give this result.

The Taylor expansion of the bias derived from (5.10) involves extremely tedious algebra. Under the assumption that f is four times continuously differentiable, it can be shown* that the approximation to the bias given by the leading nonzero term in the expansion is

$$E\hat{f}(t) - f(t) \approx \frac{h^4}{24f(t)} A(t) \int y^4 K(y) \, dy + o(h^4) \qquad (5.11)$$

---

*I am most grateful to James Davenport for helping me to use a computer algebraic manipulation package to obtain these results.

where

$$A(t) = -\frac{f^{iv}(t)}{f(t)} + \frac{8f'''(t)f'(t)}{f(t)^2} + \frac{6f''(t)^2}{f(t)^2}$$
$$-\frac{36f''(t)f'(t)^2}{f(t)^3} + \frac{24f'(t)^4}{f(t)^4} \qquad (5.12)$$

The expression (5.11) is interesting because it shows that the adaptive kernel estimator using a non-negative symmetric kernel has a bias of the same order, $O(h^4)$, as a fixed-kernel estimator with kernel satisfying $\int y^2 K(y)\,dy = 0$. In contrast to these fixed kernel estimates, the adaptive kernel estimate cannot go negative and is always a bona fide probability density. Thus the theoretical justification of the choice $\alpha = 1/2$ rests on the happy combination of this improved bias behaviour with the adaptive nature of the estimates intermediate between the fixed kernel and nearest neighbour extremes.

Practical experience reported by Abramson (1982) and others suggests that the choice $\alpha = 1/2$ gives good results and that 'the technique is no asymptotic curiosity'. Some examples will be discussed in Section 5.3.5.

5.3.4 Automatic choice of the smoothing parameter

Because of the relative insensitivity of the adaptive kernel method to the construction of the pilot estimate, the appropriate way to think about choosing the smoothing parameter automatically would appear to be to regard the λ_i as fixed and given, and then to consider the behaviour of the estimator as h varies. Of course in many applications a subjective choice of h will suffice. A natural, if ad hoc, automatic choice would be to use the same value as would be chosen using a standard density for the fixed kernel method, as in Sections 3.4.2 and 4.3.2. It should of course be pointed out that the theoretical foundation of this approach is not as strong as it would be for the fixed kernel method.

The principle of least-squares cross-validation, as discussed in Section 3.4.3, extends easily to the adaptive kernel method. Just as in (3.35), minimizing the score function

$$M_0(h) = \int \hat{f}^2 - 2n^{-1} \sum_i \hat{f}_{-i}(\mathbf{X}_i)$$

should give a choice of h that approximately minimizes the mean integrated square error. The function $\hat{f}_{-i}$, the density estimate constructed from all data points except $\mathbf{X}_i$, will be given by

$$\hat{f}_{-i}(\mathbf{t}) = (n-1)^{-1} \sum_{j \neq i} h^{-d} \lambda_j^{-d} K\{h^{-1}\lambda_j^{-1}(\mathbf{t} - \mathbf{X}_j)\}$$

and so we will have

$$n^{-1} \sum_i \hat{f}_{-i}(\mathbf{X}_i) = n^{-1}(n-1)^{-1} \sum_i \sum_{j \neq i} h^{-d} \lambda_j^{-d} K\{h^{-1}\lambda_j^{-1}(\mathbf{X}_i - \mathbf{X}_j)\}.$$

Obtaining a computationally explicit form corresponding to (3.37) for the term $\int \hat{f}^2$ is more difficult in this case and it may be best to evaluate $\int \hat{f}^2$ by numerical integration. To make some analytic progress, define $K^{(2)}(\mathbf{t}; h_1, h_2)$ to be the convolution of $h_1^{-d}K(h_1^{-1}\mathbf{t})$ and $h_2^{-d}K(h_2^{-1}\mathbf{t})$, so that

$$K^{(2)}(\mathbf{t}; h_1, h_2) = \int h_1^{-d}h_2^{-d}K\{h_1^{-1}(\mathbf{t} - \mathbf{s})\}K\{h_2^{-1}\mathbf{s}\}\,d\mathbf{s}.$$

The convolution $K^{(2)}$ is most easily expressed if K is the standard normal density, because $K^{(2)}$ will then be the normal density with variance $(h_1^2 + h_2^2)$. In terms of $K^{(2)}$, we will have

$$\int \hat{f}^2 = n^{-2} \sum_i \sum_j K^{(2)}(\mathbf{X}_i - \mathbf{X}_j; \lambda_i h, \lambda_j h). \qquad (5.13)$$

It is really only in the case of the normal kernel that (5.13) will give a useful expression for $\int \hat{f}^2$, and the normal kernel is rather inappropriate for the adaptive kernel method because of the large number of evaluations of the exponential function involved in the calculation of (5.8). For reasons of computational efficiency, one of the kernels given in (4.4), (4.5) or (4.6) is likely to be preferable.

5.3.5 *Some examples*

In this section some examples of the application of the adaptive kernel method with Abramson's choice $\alpha = 1/2$ will be presented. The first example is constructed from simulated data. Two hundred observations were simulated from the shifted log-normal density shown as the solid curve in Fig. 5.1. An ordinary kernel estimate for these data is presented as the dotted curve in Fig. 5.1. The window

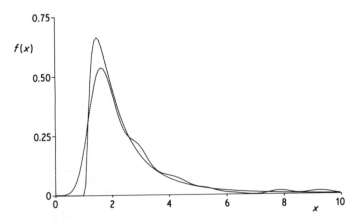

Fig. 5.1 *Solid curve: log-normal density. Dotted curve: kernel estimate for 200 simulated observations from log-normal density.*

width for this estimate has been calculated from the data using the standard formula (3.31) with a kernel scaled to have standard deviation equal to one. It can be seen from Fig. 5.1 that the fixed-width kernel estimate oversmooths the main part of the density somewhat, while there are still random fluctuations in the tails. The adaptive kernel estimator for the same data is presented in Fig. 5.2.

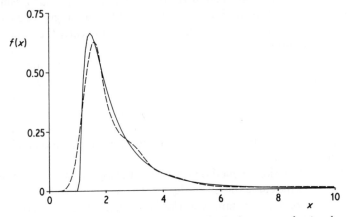

Fig. 5.2 *Solid curve: log-normal density. Dashed curve: adaptive kernel estimate for 200 simulated observations from log-normal density.*

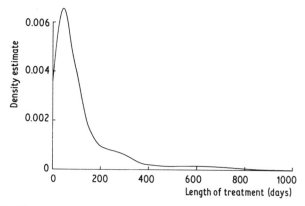

Fig. 5.3 *Adaptive kernel estimate for the suicide data discussed in Chapter 2.*

The value for the bandwidth h is the same as that used in Fig. 5.1. The performance of the estimator is greatly improved both in the tail and in the main part of the density. The bump near 3.0 is still present but is reduced in size, and the adaptive estimator copes slightly better with the steep part of the density near 1.0.

The second example, given in Fig. 5.3, is constructed from the suicide data discussed in Chapter 2. The curve given is the adaptive kernel estimate with bandwidth h again given by formula (3.31). Since the true density must be zero for negative values of time, the density estimate has been truncated at zero; an alternative approach using the reflection methods of Section 2.10 is possible and this gives a curve which is virtually constant in the range (0,50) and nearly the same as Fig. 5.3 for larger values of time. It can be seen that the estimate in Fig. 5.3 provides a good compromise between the estimates presented in Figs 2.9 and 2.11. As was pointed out in Section 5.3.2, the Breiman–Meisel–Purcell estimate presented in Fig. 2.11 corresponds to an adaptive kernel estimate with sensitivity parameter $\alpha = 1$; thus the figure provides practical support for the contention that this value of α gives an estimate whose degree of smoothing adapts too much to the pilot estimate.

The adaptive kernel method applied to the logarithms of the data points gives the curve shown in Fig. 5.4. The bump in the lower tail of the corresponding constant-width kernel estimate, Fig. 2.12, has been eliminated, and the clear indication is that the data set has a log-normal shape slightly skewed to the right.

As the final example of this section, the adaptive kernel estimate

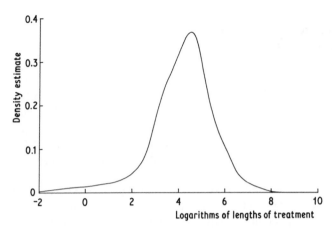

Fig. 5.4 *Adaptive kernel estimate for logarithms of suicide data.*

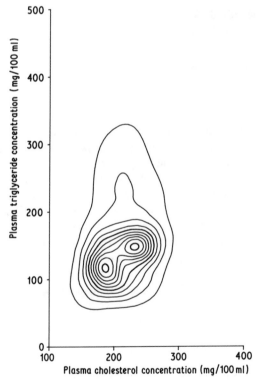

Fig. 5.5 *Adaptive kernel estimate for plasma lipid data.*

was applied to the plasma lipid data presented in Section 4.2.3. The bandwidth h is set to 40 as in Fig. 4.7, and the resulting estimate is given in Fig. 5.5. The kernel K_3 of (4.6) was used. Comparison of the two figures shows how much better behaved the adaptive kernel estimate is in the tails. The adaptive kernel estimate suggests, rather tentatively, the existence of a third cluster of cases near the point (210.0, 230.0). The contour plot in Fig. 5.5 was produced by the technique of Silverman (1981a) described at the end of Section 4.4.1, by interpolation from a grid of values and gradients of the adaptive estimate.

To sum up, the adaptive kernel method with $\alpha = 1/2$ has good practical and theoretical properties, and is well worth serious consideration if a method is required that is more accurate in the tails than the fixed-width kernel method.

5.4 Maximum penalized likelihood estimators

The maximum penalized likelihood approach, briefly outlined in Section 2.8, is an attempt to apply the ideas of maximum likelihood to curve estimation. It was demonstrated in Section 2.8 that the method of maximum likelihood is not helpful when applied directly to density estimation, since the likelihood is (effectively) maximized by the 'density' (2.7) that is a sum of delta-function spikes at the observations. However, good estimates of the density can be obtained if the likelihood is penalized by a term which takes into account the degree of roughness or local variability of the density.

In this section, only the case of univariate densities will be discussed. The extension of the penalized likelihood approach to multivariate density estimation is possible in principle, but has not been the subject of very much attention, not least because of the severe computational difficulties involved. However, there has been some work on applying the penalized likelihood approach to the related problem of nonparametric multivariate regression; see Silverman (1985a, Section 8.2 and the Discussion) for further references on this topic.

5.4.1 *Definition and general discussion*

As in Section 2.8, the penalized log likelihood of a density g given the data will be defined by

$$l_\alpha(g) = \sum_{i=1}^{n} \log g(X_i) - \alpha R(g) \qquad (5.14)$$

where the positive number α is the *smoothing parameter* and $R(g)$ is the *roughness penalty*. Notice that R is a functional, that is to say a quantity that depends on the function g; the choice of R will be discussed in detail below.

Let S be the class of all functions g that are sufficiently smooth for $R(g)$ to be defined, and that satisfy

$$\int g(x)\,dx = 1, \, g(x) \geq 0 \text{ for all } x \text{ and } R(g) < \infty. \qquad (5.15)$$

Thus S is the class of all probability densities whose roughness, measured in terms of R, is finite. A density $\hat{f}$ will be called a *maximum penalized likelihood density estimate* if

$$\hat{f} \text{ is in } S$$

and $\qquad\qquad\qquad\qquad\qquad\qquad\qquad\qquad\qquad\qquad (5.16)$

$$l_\alpha(\hat{f}) \geq l_\alpha(g) \qquad \text{for all } g \text{ in } S,$$

in other words if $\hat{f}$ maximizes $l_\alpha(g)$ over the class of all g satisfying (5.15).

The penalized likelihood approach is appealing for philosophical reasons. It makes explicit the notion that there are two often conflicting aims in curve estimation: one is to maximize fidelity to the data, as measured by the log likelihood $\Sigma \log g(X_i)$, while the other is to avoid curves which exhibit too much roughness or rapid variation, as measured by $R(g)$. The choice of the smoothing parameter controls the balance between smoothness and goodness-of-fit, while the choice of the roughness penalty determines exactly what kind of behaviour in the density estimate is considered to be undesirable in excess. For example, the choice $R(g) = \int g''^2$ will have high value if g exhibits a large amount of local curvature, and the value zero if g is a straight line. Unfortunately, the implicit nature of the definition of the estimate $\hat{f}$ itself, as the solution of a maximization problem, is the price to be paid for the explicit statement of the aims of the estimation.

Another philosophically attractive feature of penalized likelihood is that it places density estimation within the context of a unified approach to curve estimation; see Section 5.4.5, where some Bayesian connotations of the method will also be discussed.

5.4.2 *The approach of Good and Gaskins*

The first authors explicitly to apply the penalized likelihood approach to density estimation were Good and Gaskins (1971); they refer to other work published by Good. They based their roughness penalty on the square root of f; letting $\gamma = \sqrt{f}$, they first considered the roughness penalty.

$$R(f) = \int \gamma'^2. \tag{5.17}$$

By simple calculus it is easily shown from (5.17) that

$$4R(f) = \int f'^2/f, \tag{5.18}$$

the Fisher information for the location parameter θ in the family $f(\cdot - \theta)$ (see Example 4.14 of Cox and Hinkley, 1974).

The advantage of working with γ rather than f is that the constraint $f(x) \geq 0$ will automatically be satisfied if γ is real; furthermore, the constraint $\int f = 1$ will be replaced by $\int \gamma^2 = 1$, an easier constraint under the numerical method used by Good and Gaskins (1971) and described briefly below.

The penalty (5.17) penalizes for slope rather than curvature in the estimates; to penalize for curvature, Good and Gaskins (1971) suggest using the penalty

$$R(f) = \int \gamma''^2 \tag{5.19}$$

or even a linear combination of (5.17) and (5.19).

In their original paper and in subsequent work, Good and Gaskins (1971, 1980) develop a computational approach to the calculation of maximum penalized likelihood density estimates with roughness penalty (5.17) or (5.19). The basic idea is to expand γ as an orthogonal series

$$\gamma(x) = \sum \gamma_m \phi_m(x) \tag{5.20}$$

where the series $\phi_m(x)$ is some suitable series of orthogonal functions, such as the sine and cosine functions that would make (5.20) a Fourier expansion. The expansion (5.20) is pursued to a large number of terms. The maximum penalized likelihood is then expressed, by substituting (5.20), in terms of the coefficients γ_m; maximizing this function gives values $\hat{\gamma}_m$ which are substituted back into (5.20) to give $\hat{\gamma}(x)$ and hence the density estimate $\hat{f}(x) = \hat{\gamma}(x)^2$. The reader is referred to the original papers for full details of the method, including hints on the appropriate choice of orthogonal series and strategies for choosing the number of terms in the expansion (5.20). The approach is a development of the Rayleigh–Ritz method for nonlinear maximization described, for example, in Courant and Hilbert (1953, pp. 175–6).

A special method which can be used only for the roughness penalty (5.17) is developed by Ghorai and Rubin (1979). It can be show that the square root $\hat{\gamma}$ of the density estimate $\hat{f}$ satisfies

$$\hat{\gamma}(x) = \sum_{j=1}^{n} \theta_j \exp\left\{-\lambda|x - X_j|\right\} \qquad (5.21)$$

where the constants θ_j and λ are found as solutions to $(n+1)$ nonlinear equations derived from

$$\int \hat{\gamma}(x)^2 \, dx = 1 \qquad (5.22)$$

and

$$2\theta_j \lambda \alpha \hat{\gamma}(X_j) = 1 \qquad \text{for } j = 1, \ldots, n. \qquad (5.23)$$

Ghorai and Rubin (1979) give two algorithms for solving these equations to find the density estimate. The form (5.21) is a little worrying, since it shows that the derivative of $\hat{f}$ will be discontinuous at each data point, and so the density estimate will not be a smooth curve. This is a somewhat unattractive feature of the roughness penalty (5.17) and illustrates why a penalty like (5.19) is likely to be preferred. An example of a density estimate constructed using the roughness penalty (5.17) is given in Fig. 5.6. This figure is a recomputed version of Fig. 2 of Ghorai and Rubin (1979) and is based on a set of 13 data points given in that paper. In fact the kinks in the curve at points where the derivative is discontinuous are fairly clearly visible, and so the original data set can be reconstructed from the figure. Though this behaviour is not disastrous, it does demonstrate an uncomfortable feature of this particular roughness penalty.

Finally, it should be pointed out that the use of the roughness penalty (5.19) leads to some technical difficulties concerning the uniqueness and definition of the estimates. These are discussed by de Montricher, Tapia and Thompson (1975).

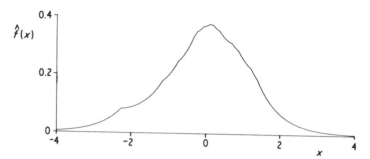

Fig. 5.6 *Maximum penalized likelihood estimate with penalty* $\int \gamma'^2$. *After Ghorai and Rubin (1979).*

5.4.3 *The discrete penalized likelihood approach*

In order to avoid some of the numerical and mathematical difficulties involved in finding the maximum penalized likelihood estimators discussed in the last subsection, Scott, Tapia and Thompson (1980) suggested replacing roughness penalties defined in terms of integrals and derivatives by approximations based on sums and differences. We shall give details of the same examples as they did; however, the approach is quite general and can easily be extended to cover other roughness penalties.

Restrict attention to a finite interval $[a, b]$ which is such that the unknown density f can be considered to be concentrated on $[a, b]$, and furthermore it can be assumed that $f(a) = f(b) = 0$. In practice $[a, b]$ would be an interval somewhat larger than the range of all the observations. Choose a moderately sized number m (typically $m = 40$) and cover $[a, b]$ with a regularly spaced mesh of points

$$a = t_0 < t_1 < \cdots < t_m = b$$

so that the mesh interval δ is defined by

$$\delta = t_j - t_{j-1} = (b - a)/m$$

Let P_m be the space of all continuous piecewise linear functions p on $[a, b]$ which are linear on each interval $[t_j, t_{j+1}]$ and which satisfy $p(a) = p(b) = 0$. Write $p_j = p(t_j)$ and define a roughness penalty

$$R(p) = \delta \sum_{j=1}^{m} \frac{(p_j - p_{j-1})^2}{\delta^2} \qquad (5.24)$$

The penalty (5.24) is clearly a discrete approximation to the integral penalty $\int p'^2$, and for functions in P_m the two penalties are in fact identical. The penalized likelihood (5.14) can be written indirectly in terms of the $(m-1)$ parameters $p_1, \ldots, p_{m-1}$ as

$$\sum_{i=1}^{n} \log p(X_i) - \alpha \delta^{-1} \sum_{j=1}^{m} (p_j - p_{j-1})^2; \qquad (5.25)$$

each $p(X_i)$ is just a linear combination of the two nearest $p(t_j)$. The discrete maximum penalized likelihood estimate $\hat{p}$ is then found by maximizing (5.25) over p in P_m subject to the constraints

$$\left. \begin{array}{ll} p_j \geq 0 \quad \text{all } j, \quad p_0 = p_m = 0 \\[2mm] \delta \sum_{j=1}^{m} p_j = 1, \end{array} \right\} \qquad (5.26)$$

and

which ensure that $\hat{p}$ is a bona fide probability density function.

Scott, Tapia and Thompson (1980) give theoretical justification for the intuitive property that $\hat{p}$ will be a good approximation to the exact maximum penalized likelihood estimator with penalty $R(f) = \int f'^2$. They also give further theoretical and practical details. In their practical work they use a penalty constructed using second differences in order to approximate the penalty $\int f''^2$. This leads to the optimization problem

$$\max \sum_{i=1}^{n} \log p(X_i) - \alpha \delta \sum_{j=1}^{m} (p_{j+1} - 2p_j + p_{j-1})^2 / \delta^4$$

subject to the constraints (5.26) and in addition setting $p_{-1} = p_{m+1} = 0$. Some examples of estimates constructed by their method are given in Tapia and Thompson (1978, Chapter 5), and a computer implementation NDMPLE of the method is included in the fairly widely available IMSL program library.

To sum up, the basic idea of the method is to reduce the potentially infinite-dimensional problem of maximizing the penalized log likelihood to an optimization problem in a fairly large but finite number of dimensions, by using a discretized form of the roughness penalty and restricting attention to a suitable class of functions p.

5.4.4 Penalizing the logarithm of the density

There are certain potential advantages in using a roughness penalty based on the logarithm of the density, and these will be discussed in this section. For further details and results see Silverman (1982b), where the estimators of this section were originally discussed.

Consider the roughness penalty

$$R(f) = \int \{(d/dx)^3 \log f(x)\}^2 \, dx \qquad (5.27)$$

When expressed in terms of the logarithm of f, the problem of finding the maximum penalized likelihood density estimate becomes, setting $g(x) = \log f(x)$,

$$\max \sum g(X_i) - \alpha \int (g''')^2 \qquad (5.28)$$

subject to

$$\int \exp\{g(x)\} \, dx = 1. \qquad (5.29)$$

Working with the logarithm of f has the advantage of eliminating the necessity for a positivity constraint on f (since $f = \exp(g)$ will automatically be positive) and reducing the quantity to be maximized in (5.28) to a quadratic form. The cost to be paid is the awkward nonlinear nature of the constraint (5.29).

The rather strange-looking roughness penalty (5.27) has the important property that it is zero if and only if f is a normal density. Thus normal densities are considered by the method to be 'infinitely smooth' because they are not penalized at all in (5.28) and so cost nothing in terms of roughness. It can be shown, in a sense made clear in Silverman (1982b, Theorem 2.1), that as the smoothing parameter α tends to infinity, the limiting estimate will be the normal density with the same mean and variance as the data. As α varies, the method will give a range of estimates from the 'infinitely rough' sum of delta functions at the data points to the 'infinitely smooth' maximum likelihood normal fit to the data. Since one of the objects of nonparametric methods is to investigate the effect of relaxing parametric assumptions, it seems sensible that the limiting case of a nonparametric density estimate should be a natural parametric estimate.

It is possible to define other roughness penalties according to other perceptions of 'infinitely smooth' exponential families of densities. The key property is that $R(f)$ should be zero if and only if f is in the required family. For example, when working on the half-line

$(0, \infty)$ a natural penalty is $\int_0^\infty \{(\log f)''\}^2$, which gives zero roughness to the exponential densities $\lambda e^{-\lambda x}$.

Detailed numerical work on the density estimates of this section remains to be done. It can be shown (see Theorem 3.1 of Silverman, 1982b) that the maximum of (5.28) subject to the constraint (5.29) can be found as the *unconstrained* maximum of the strictly concave functional

$$\sum g(X_i) - \alpha \int (g''')^2 - n \int \exp(g). \tag{5.30}$$

The form (5.30) is remarkable in that it contains no unknown Lagrange multipliers to be determined; its maximum will automatically satisfy the constraint (5.29). The fact that the estimates can be found as the unconstrained maximum of a concave functional also makes it possible to derive various theoretical properties of the estimates; see the later sections of Silverman (1982b). In particular, it can be shown that, by suitable choice of roughness penalty and under suitable smoothness conditions on the density, the estimators $\hat{f}$ achieve the bias reduction properties of Section 3.6 while still being non-negative functions that integrate to unity.

A heuristic argument given in Silverman (1984a, Section 7) demonstrates a connection between the estimates of this section and the adaptive kernel estimates discussed in Section 5.3 above. Suppose a roughness penalty of the form

$$R(f) = \int (g^{(r)} + \text{terms in lower derivatives of } g)^2$$

is used, where $g = \log f$. It turns out that the resulting maximum penalized likelihood density estimator is approximately an adaptive kernel estimator with sensitivity parameter $1/2r$. Though this result requires further theoretical and practical investigation, it is interesting because it suggests that maximum penalized likelihood estimators may provide a one-step procedure that adapts well in the tails of the density without relying on a pilot estimator.

5.4.5 Penalized likelihood as a unified approach

An interesting feature of the penalized likelihood approach is that it can be applied, in principle at least, to a wide variety of curve estimation problems. Given any curve estimation problem where the log likelihood of a putative curve g can be expressed, this log likelihood can be penalized by a roughness penalty term.

Consider, for example, the problem of nonparametric regression where observations Y_i are taken at points t_i and are assumed to satisfy

$$Y_i = g(t_i) + \varepsilon_i. \tag{5.31}$$

Here the ε_i are independent $N(0, \sigma^2)$ errors and g is an unknown curve to be estimated. The log likelihood of g in the model (5.31) is, up to a constant,

$$-\frac{1}{2\sigma^2} \sum \{Y_i - g(t_i)\}^2. \tag{5.32}$$

The quantity (5.32) will be maximized by any curve g that actually interpolates the data points, and such a curve will usually be unacceptable because it displays too much rapid fluctuation.

Using the roughness penalty $\int g''^2$, the penalized log likelihood becomes

$$-\frac{1}{2\sigma^2} \sum \{Y_i - g(t_i)\}^2 - \alpha \int g''^2. \tag{5.33}$$

It turns out that (5.33) can be maximized far more easily than the penalized log likelihood for density estimation. Just as in density estimation, the smoothing parameter α controls the trade-off between goodness-of-fit and smoothness; the degree of smoothness increases as α increases. The solution to the maximization problem is a cubic spline, i.e. a piecewise cubic polynomial with continuous second derivative, and the method of nonparametric regression via the penalized log likelihood (5.33) is often called *spline smoothing*. A detailed discussion is beyond the scope of this book, but for further details the reader is referred, for example, to Wegman and Wright (1983) and Silverman (1985a).

In some cases the likelihood itself is not available but an approach via partial or conditional likelihood is possible. For example, Silverman (1978c) considered the problem of estimating the logarithm of the ratio of two densities f and g given samples from both of them. This problem is related to discriminant analysis and will be discussed further in Chapter 6. Let $\beta(x) = \log \{f(x)/g(x)\}$ and suppose that the observations are (Z_i, ε_i), where Z_i is the position of the ith observation and ε_i is equal to 1 if the observation is drawn from f

and 0 if it is drawn from g. A standard conditional likelihood argument gives the conditional log likelihood

$$l(\beta) = \sum (\varepsilon_i \beta(Z_i) - \log[1 + \exp\{\beta(Z_i)\}]). \tag{5.34}$$

The expression (5.34) is maximized by setting $\beta(Z_i) = +\infty$ if $\varepsilon_i = 1$ and $\beta(Z_i) = -\infty$ if $\varepsilon_i = 0$. Much better estimates of the log density ratio are obtained by introducing a roughness penalty term and then maximizing. Defining $l(\beta)$ as in (5.34), the quantity to be maximized becomes

$$l(\beta) - \alpha \int \beta''^2 \tag{5.35}$$

where, as usual, α is a smoothing parameter. As in the nonparametric regression example, the maximizer of (5.35) is a cubic spline; its computation in practice is somewhat more difficult. For further details and an application, see Silverman (1978c). References to other uses of penalized likelihood in curve estimation are given by Silverman (1985b).

In general, penalized likelihood can be given something of a Bayesian interpretation. The basic idea, suggested by Good and Gaskins (1971), is to place an improper prior distribution over the space of all smooth curves. The idea of viewing nonparametric curve estimation in a Bayesian context dates back at least to Whittle (1958). (See also Leonard, 1978.) The prior density of a curve f is taken to be proportional to $\exp\{-\alpha R(f)\}$, so that the smoothing parameter α is a parameter of the prior. The penalized log likelihood can then be shown to correspond, up to a constant, to the logarithm of the posterior density, and so the maximum penalized likelihood estimate $\hat{f}$ is the mode of the posterior density over the space of curves. Unfortunately, there are certain drawbacks to this approach, caused in part at least by the infinite-dimensional nature of the space of smooth curves. Although the intention is to choose among curves for which the roughness $R(f)$ is finite, the posterior distribution is entirely concentrated *outside* the space of such smooth curves; for further details and a possible resolution of this paradox see Silverman (1985a, Section 6.1). Despite difficulties of this kind, it is very natural to think of curve estimation in a Bayesian way, because the choice of smoothing parameter is, effectively, a choice of how smooth the statistician thinks the curve is likely to be.

CHAPTER 6

Density estimation in action

In this chapter, various contexts in which density estimates have been used will be described. The survey is not intended to be exhaustive or prescriptive, but it is hoped that it will emphasize the applicability of density estimation to a variety of topics. With this aim in mind, we shall focus attention on methods based fairly explicitly on density estimation ideas, without necessarily claiming that these are the only or even the best approaches.

6.1 Nonparametric discriminant analysis

It was pointed out in Chapter 1 that the original intended use of density estimation (Fix and Hodges, 1951) was as part of a proposed nonparametric version of discriminant analysis. In this section the application of density estimation to statistical discrimination will be discussed.

The basic problem of discrimination is easily stated. We are given a sample $\mathbf{X}_1, \ldots, \mathbf{X}_n$ known to come from a population A, and $\mathbf{Y}_1, \ldots, \mathbf{Y}_m$ known to come from a population B. These samples form the *training set*. Given a new observation $\mathbf{Z}$, does $\mathbf{Z}$ come from population A or from population B?

A typical context for the discrimination problem is automatic medical diagnosis. Each training observation $\mathbf{X}_i$ or $\mathbf{Y}_j$ consists of a vector of medical test results and data for a patient whose diagnosis is known; population A consists of those patients who have a particular disease while population B consists of those patients who do not. The new observation $\mathbf{Z}$ is then a data vector on a patient whose diagnosis is unknown. In the medical context the data vectors $\mathbf{X}_1, \ldots, \mathbf{X}_n$, $\mathbf{Y}_1, \ldots, \mathbf{Y}_m$, and $\mathbf{Z}$ are usually fairly high-dimensional multivariate observations consisting of some categorical variables and some continuous variables.

Another extremely important context is remote sensing satellite data. Here it is of interest to assign small regions (pixels) on the Earth's surface to one of a number of classes (e.g. types of crop) on the basis of satellite observations. The observation collected for each pixel is typically a vector giving the intensities of reflected radiation in four spectral bands, thus giving a four-dimensional measurement somewhat related to the colour of the pixel. For some pixels, surveys are done on the ground to discover in which class they actually fall, and these pixels form the training set; it is then necessary to classify the remaining pixels (the vast majority of the data) on the basis of the data observed by the satellite. There are, of course, numerous other allocation and discrimination problems that arise in practice, in fields ranging from social science to engineering; see Hand (1981) for a review of some of these.

6.1.1 *Classical approaches*

In order to place the nonparametric approach in context, the classical approach to discriminant analysis will be reviewed very briefly. For further details of the arguments underlying this section see, for example, Mardia, Kent and Bibby (1979, Chapter 11). In this section and much of what follows, we shall assume that the observations $X_1, \ldots, X_n$, $Y_1, \ldots, Y_m$, and Z are all continuous multivariate observations.

Suppose, first of all, that observations drawn from population A have known probability density function f_A while those from population B have known probability density function f_B. Then an approach based on maximum likelihood would allocate Z to population A if

$$f_A(Z) \geqslant f_B(Z); \tag{6.1}$$

a more general hypothesis testing or Bayesian approach allocates Z to population A if

$$f_A(Z) \geqslant c f_B(Z), \tag{6.2}$$

where the constant c is chosen by reference to the probabilities of misclassification or, if possible, by considering the prior probability that Z comes from population A and the various utilities of correct and incorrect classification. If prior odds that Z comes from population A can be assessed, then posterior odds ($\mathbf{Z}$ is from A) $= \dfrac{f_A(\mathbf{Z})}{f_B(\mathbf{Z})} \times$ prior odds ($\mathbf{Z}$ is from A)

by a standard Bayesian calculation.

In most practical problems, the densities f_A and f_B cannot be assumed to be known, and so the discriminant rule must be estimated from the training set in some way. One natural approach is to suppose that the unknown densities come from some parametric family and to estimate the parameters underlying each of the populations A and B from the training set. For example, the distributions of the two populations might be taken to be multivariate normal with respective mean vectors μ_A and μ_B and common variance matrix. The means and the variance can be estimated by the sample means $\bar{\mathbf{X}}$ and $\bar{\mathbf{Y}}$ of the two training populations and the pooled sample covariance matrix $S = (m + n - 2)^{-1} \{ \Sigma_i (\mathbf{X}_i - \bar{\mathbf{X}})(\mathbf{X}_i - \bar{\mathbf{X}})^{\mathrm{T}} + \Sigma_j (\mathbf{Y}_j - \bar{\mathbf{Y}})(\mathbf{Y}_j - \bar{\mathbf{Y}})^{\mathrm{T}} \}$. With these estimates inserted to give parametric estimates of f_A and f_B, the rule (6.1), in this special case, becomes a *linear discriminant rule*: allocate $\mathbf{Z}$ to population A if

$$\{ \mathbf{Z} - \tfrac{1}{2}(\bar{\mathbf{X}} + \bar{\mathbf{Y}}) \}^{\mathrm{T}} S^{-1} (\bar{\mathbf{X}} - \bar{\mathbf{Y}}) \geq 0. \tag{6.3}$$

This is Fisher's (1936) linear discriminant rule, though Fisher did not arrive at it by the approach outlined above, but by using an argument based on an index of separation of the two populations.

The parametric approach to discriminant analysis can obviously be extended. For example, the populations A and B might still be assumed to be multivariate normal, but not necessarily with equal variance matrices. In this case estimation of the means and co-variances from the training set and substitution into (6.1) or (6.2) will lead to a *quadratic discriminant rule* where the allocation rule depends on the value of a quadratic form in the observed data vector $\mathbf{Z}$.

6.1.2 The nonparametric approach

Fix and Hodges (1951) very naturally went on from the parametric case of discriminant analysis to ask how to proceed if nothing is known about the densities f_A and f_B, 'except possibly for assumptions about existence of densities, etc.' Their suggested approach is to estimate the densities f_A and f_B from the training set using nonparametric density estimates, and to substitute these density estimates into (6.1) or (6.2) to give a nonparametric discriminant rule.

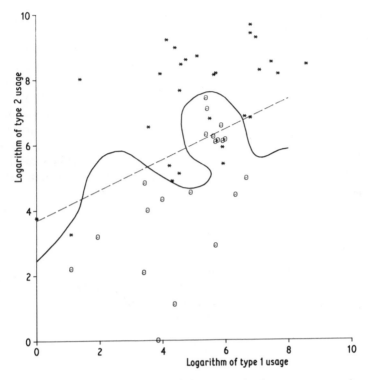

Fig. 6.1 *Nonparametric and linear discriminant rules for computer user data.*

In order to carry out this procedure, it is necessary to settle on the method of density estimation to be used and to specify the smoothing parameters; these questions will be discussed in Section 6.1.3.

An example (really intended only for illustration) is given in Fig. 6.1. The data used to construct this figure were taken from Hand (1981, Table 2.1). The variables measured were the amounts of computer usage, under each of two operating systems, for institutional users of the University of London Computer Centre. The institutions were divided into two classes, medical and non-medical, and log transformations of the data were taken. Only institutions with nonzero usage of both kinds were considered. The solid curve in Fig. 6.1 shows the boundary of the nonparametric allocation rule obtained by constructing adaptive kernel estimates for each of the two populations and substituting these into (6.1). (The adaptive kernel estimates were calculated as in Section 5.3, setting $\alpha = 0.5$ and $h = 2.0$, using the kernel (4.6) and using a kernel estimate with $h = 2.0$ as the pilot estimate.) For comparison, the dashed line shows the linear discriminant rule (6.3). The most important qualitative difference between the rules is that the nonparametric rule takes clear account of the cluster of medical schools near the point (6.0, 6.0), while the linear rule does not. The observed distribution of medical schools is somewhat non-normal because of this cluster, so it is not surprising that the linear rule does not perform well; the nonparametric rule is able to adapt to this feature in the data. The discrepancies between the rules around (3.0, 5.0) are less important since this is a region very sparse in data.

A quantitative assessment of the merits of the two rules can be obtained by considering the *leaving-one-out misclassification score*, calculated as follows: leave each data point out in turn, and recalculate the discriminant rules from the remaining data; then classify the omitted data point, and count the number of cases in which the point is allocated to the wrong population. On this basis the linear discriminant rule misclassified 12 out of the 49 points, while the nonparametric rule misclassified only 7 of the 49 points.

6.1.3 Operational details

It is extremely important to keep in mind that in the present context density estimates are used in order to give a discrimination rule rather than as an end in themselves. Good estimates of f_A and f_B will give a good discrimination rule, since knowing f_A and f_B exactly would (by standard statistical arguments) give the best possible information about which population unclassified observations come from. However, it may well be the case that even quite crude density estimates will give reasonable results.

The discussion in earlier chapters of this book suggests using the kernel or adaptive kernel method for estimating the unknown densities f_A and f_B. The choice of smoothing parameter can be made either by one of the methods discussed previously or by reference to the particular aims of the classification problem in hand. For example, one possible approach is to use the same smoothing parameter h for both populations, and to assess various values of h by considering the leaving-one-out resubstitution error rate. The disadvantages of such an approach are that it is only really practicable when a single smoothing parameter is used, and that it may well be appropriate to use different smoothing parameters for the different populations.

Several authors have conducted simulation studies to investigate the density estimation approach to nonparametric discriminant analysis, and to compare it with classical approaches. However, much work remains to be done on this topic and any results must be regarded as tentative. Perhaps the most detailed study has been reported by Remme, Habbema and Hermans (1980). Their nonparametric method uses separate kernel estimates for each population in the training set, with smoothing parameters chosen by the likelihood cross-validation method described in Section 3.4.4. They considered four different models for the underlying data and compared their kernel method with linear and quadratic discriminants. The following conclusions emerge from their study:

(a) Not surprisingly, for normal populations with equal variances the linear discriminant performed best, but even in this case the kernel method performed quite well, particularly when the populations were not very well separated.
(b) For all the other cases considered, the kernel method came out best, or very nearly so. The kernel method performed particularly well on non-normal data with reasonably large training sets (more than 50 observations).
(c) For long-tailed distributions, the behaviour of the kernel method (though superior to that of the linear discriminant) left something to be desired, and in particular the method of likelihood cross-validation gave a rather inappropriate choice of smoothing parameter for the discriminant problem. For short-tailed distributions, likelihood cross-validation gave very good results.

The final conclusion of Remme, Habbema and Hermans (1980) is that 'although each of the methods yields the best results in some situations, the kernel method is the only one which was the best or close to the best in all situations we considered. In particular the present practice of the nearly exclusive use of linear discriminant analysis cannot be justified by our results.'

Another relevant study by the same authors (Habbema, Hermans and Remme, 1978) compares the kernel method with a variable kernel approach using the method of Breiman, Meisel and Purcell (1977) described in Section 5.3.2 above. In both cases likelihood cross-validation was used to choose the smoothing parameter h. The variable kernel approach gave improved results for the long-tailed models, while giving results comparable to the kernel method for the other models considered. Although its use has not been studied carefully in this context, the discussion of Section 5.3 indicates that the adaptive kernel method with $\alpha = \frac{1}{2}$ should give good results. The relatively disappointing behaviour of likelihood cross-validation for long-tailed distributions is not very surprising in view of the difficulties mentioned in Section 3.4.4.

6.1.4 Generalizations for discrete and mixed data

In many of the real applications of discriminant analysis, some or all of the components of the observed data vectors are discrete or categorical observations. Nonparametric discrimination as described above can be extended to deal with such data, by generalizing the ideas of density estimation to deal with discrete and mixed data.

Consider the special case of multivariate binary data, where every observed variable effectively takes the value 0 or 1. The distribution of a random multivariate binary vector of length k is given by the probabilities of each of the 2^k possible outcomes. In real problems, k typically falls in the range 5 to 20, and the training sets are far too small for there to be any hope of estimating the probability of a particular outcome by the proportion of times it occurs in the training set. Let B^k be the space $\{0, 1\}^k$ of possible multivariate binary observations. The ideas of kernel density estimation can be extended to provide a method for smoothing sets of observations on B^k in order to estimate their underlying distributions. Given vectors $\mathbf{x}$ and $\mathbf{y}$ in B^k, let $d(\mathbf{x}, \mathbf{y})$ be the number of disagreements in corresponding components of $\mathbf{x}$ and $\mathbf{y}$; it is easy to see that

$$d(\mathbf{x}, \mathbf{y}) = (\mathbf{x} - \mathbf{y})^{\mathrm{T}}(\mathbf{x} - \mathbf{y}). \tag{6.4}$$

Now, for any λ with $\frac{1}{2} \leqslant \lambda \leqslant 1$, define a kernel K by

$$K(\mathbf{y}|\mathbf{x}, \lambda) = \lambda^{k-d(\mathbf{x}, \mathbf{y})}(1 - \lambda)^{d(\mathbf{x}, \mathbf{y})}. \tag{6.5}$$

The kernel K can be shown to satisfy

$$\sum_{\mathbf{y}} K(\mathbf{y}|\mathbf{x}, \lambda) = 1 \quad \text{for all } \mathbf{x} \text{ and } \lambda. \tag{6.6}$$

Given a sample $\mathbf{X}_1, \ldots, \mathbf{X}_n$ of observations drawn from a distribution p

on B^k, the kernel estimate $\hat{p}$ of p is defined by

$$\hat{p}(\mathbf{y}) = n^{-1} \sum_i K(\mathbf{y}|\mathbf{X}_i, \lambda). \tag{6.7}$$

The parameter λ governs the degree of smoothing applied to determine the estimate. When $\lambda = 1$, the whole weight of $K(\mathbf{y}|\mathbf{x}, \lambda)$ is concentrated at $\mathbf{y} = \mathbf{x}$, and so $\hat{p}(\mathbf{y})$ is just the proportion of the data set for which $\mathbf{X}_i = \mathbf{y}$. On the other hand, when $\lambda = \frac{1}{2}$, $K(\mathbf{y}|\mathbf{x}, \lambda)$ gives the same weight $(\frac{1}{2})^k$ to every $\mathbf{y}$ in B^k, regardless of the observation $\mathbf{x}$, and so $\hat{p}(\mathbf{y})$ is the uniform distribution over B^k. The method of likelihood cross-validation is easily generalized to give an automatic choice of λ; the score to be maximized is

$$\sum \log \hat{p}_{-i}(\mathbf{X}_i) \tag{6.8}$$

where $\hat{p}_{-i}$ is the estimate of p obtained from all the data points except $\mathbf{X}_i$.

Given a training set of observations $\mathbf{X}_i$ from population A and $\mathbf{Y}_j$ from population B, estimates $\hat{p}_A$ and $\hat{p}_B$ of the underlying distributions can be obtained by this approach. A new observation $\mathbf{Z}$ can be classified by consideration of $\hat{p}_A(\mathbf{Z})/\hat{p}_B(\mathbf{Z})$ just as $f_A(\mathbf{Z})/f_B(\mathbf{Z})$ was used in Section 6.1.1. Aitchison and Aitken (1976), in addition to providing the material on which the above discussion is based, have applied the method to the analysis of some data on the diagnosis of the disease *Keratoconjunctivitis sicca* (KCS) reported by Anderson *et al.* (1972). For each patient in the original study, the presence or absence of each of 10 features was determined, making up a multivariate binary vector of length $k = 10$. The training set consisted of 40 patients known to have the disease and 37 known to be free of it. Likelihood cross-validation was used to give choices of the smoothing parameters ($\lambda = 0.843$ for the KCS patients and 0.960 for the others). The resulting nonparametric discrimination rule was used to classify a test set of 41 new patients of whom 24 in fact had the disease; the discriminant rule correctly diagnosed all 41 test patients on the basis of the 10 observed features.

The ideas of this section can be applied to more general types of data. The basic technique is to define a suitable version of the kernel method which is used to smooth the observed data, giving an estimate of probability density or mass over the appropriate space. For example, if the observations have k_1 binary components and k_2

continuous components, a possible kernel might be

$$K(\mathbf{y}|\mathbf{x},\lambda,h) = \lambda^{k_1 - d_1(\mathbf{x},\mathbf{y})}(1-\lambda)^{d_1(\mathbf{x},\mathbf{y})}h^{-k_2}\phi\{h^{-1}d_2(\mathbf{x},\mathbf{y})\}(2\pi)^{\frac{1}{2}(1-k_2)} \quad (6.9)$$

where d_1 is the distance (6.4) between the binary components, d_2 is the Euclidean distance between the continuous components, ϕ is the normal density function, and λ and h are smoothing parameters. If $\mathscr{S}$ is the space of possible observations, then

$$\hat{f}(\mathbf{y}) = n^{-1}\sum_i K(\mathbf{y}|\mathbf{X}_i,\lambda,h) \qquad \text{for } \mathbf{y} \text{ in } \mathscr{S} \quad (6.10)$$

gives an estimate of the density underlying the observations $\mathbf{X}_i$. For possible extensions of the weight function (6.9) to deal with ordered and unordered categorical variables, and also missing data, see, for example, Aitchison and Aitken (1976), Titterington *et al.* (1981), Hermans *et al.* (1982), and Habbema, Hermans and Remme (1978). The last of these papers also considers a variable kernel approach, where the smoothing parameter in the estimate (6.10) is allowed to vary between data points on the basis of a pilot estimate.

An extremely interesting comparative study of various discriminant methods on a real data set was carried out by Titterington *et al.* (1981). They considered a set of 1000 patients with severe head injury. For each patient, up to 12 categorical features of different types were observed shortly after injury. It was of interest to assess prognosis on the basis of these features, and the patients were subsequently classified according to their degree of eventual recovery. Perhaps because of the high dimensionality and the large number of missing observations, the kernel methods were a little disappointing in their performance. However, in their discussion the authors express the belief that kernel methods will come into their own with more data and further methodological development.

An interesting possibility in nonparametric discrimination using density estimates is the definition of an *atypicality index* for observations assigned to a particular population. This gives an answer to the question, 'How typical a representative of the population is this observation?' We would, quite rightly, treat with some care an observation $\mathbf{z}$ that was allocated to a population but which was an unusual member of that population. Atypicality indices were discussed by Aitchison and Aitken (1976) in the discrete data context. If $\hat{f}$ is the estimated density underlying the population, then their definition of the atypicality of $\mathbf{z}$ as a member of the population is the probability that a randomly chosen ζ from the population has $f(\zeta)$ greater than $\hat{f}(\mathbf{z})$. Atypicality indices can be calculated either by simulation or, in the discrete case, by exhaustive enumeration. They certainly deserve more attention and investigation than they have so far received.

6.1.5 Discussion and conclusions

It is difficult, in the present state of knowledge, to give a fair assessment of the usefulness of the nonparametric discrimination techniques described here. Theoretical properties of these techniques are discussed by Prakasa Rao (1983, Chapter 8), who displays their asymptotic optimality in a certain sense. Loosely speaking, the idea of this theoretical work is that, if large samples are available, consistent estimates of the underlying densities can be obtained and therefore, under very mild conditions, the eventual behaviour of the nonparametric rules will be as good as if the population densities were actually known. The practical difficulty of this argument is that the sample sizes actually required may be extremely large, especially if the data are of high dimensionality. Nevertheless, it gives mathematical weight to the intuitive feeling that the nonparametric methods are well worth considering, especially when extensive data sets are collected.

Practical experience with the methods suggests that there may be some scope for improvement but that even in their present state they can give good results. The computer package ALLOC80 (see Hermans et al., 1982) is based on density estimation ideas and is in quite wide practical use. The package incorporates almost all of the features discussed in this section, including both the fixed and variable kernel approaches, and the capability of dealing with mixed and discrete data of various kinds. It also provides a method of variable selection, a useful feature for high-dimensional data.

A much fuller description both of the discrimination problem in general and of nonparametric discrimination in particular is given by Hand (1981), to which the reader is referred for additional material and references not included above. The emphasis in this section has been on nonparametric discrimination using density estimation. However, this is only one of a number of alternatives to the parametric approach; other nonparametric methods include recursive partitioning (Breiman et al., 1984) and nearest neighbour approaches (Cover and Hart, 1967; Hellman, 1970). The monograph by Hand (1982) gives a specialist treatment of the kernel density estimation approach. Developments in knowledge-based computer 'expert systems' and the growth of discriminant problems involving very large data sets will, it is hoped, stimulate further theoretical and practical work in this important area.

6.2 Cluster analysis

The main aim of cluster analysis is to divide a given population into a number of clusters or classes. In contrast to discriminant analysis, no previous information is available about the properties – or even the existence – of the various groups; the number of groups and the rules of assignment into these groups have to be discerned solely from the given data, without reference to a training set. Cluster analysis is called *classification* by some authors; unfortunately the word 'classification' is also often used to cover a variety of activities including discriminant analysis.

A careful and thorough recent review of cluster analysis is provided by Gordon (1981). Another excellent, and more provocative, survey is provided by Cormack (1970), whose paper begins as follows: 'The availability of computer packages of classification techniques has led to the waste of more valuable scientific time than any other "statistical" innovation (with the possible exception of multiple regression techniques)'. Cormack lays much emphasis on the importance of pausing to ask 'why and when as well as how' cluster analysis should be performed. Nevertheless, there are many fields of investigation where cluster analysis, used carefully, has yielded valuable results, for example in archaeology and in the biological sciences. See the references given by Gordon (1981).

The discussion in this section will be confined to clustering methods that make use of density estimation. There are probably hundreds of clustering methods available in the literature, and little attempt will be made to compare the methods based on density estimation with other methods. All that will be done here is to give the general flavour of the way that density estimation can be used in cluster analysis.

Suppose that each object i in the set to be clustered can be represented as a point X_i in d-dimensional space. In much of what follows no distinction will be made between the object i and the corresponding point X_i. Suppose there are n objects to be clustered.

All the methods discussed in this section rest on the same basic idea, that clusters in the set $\mathbf{X}_1, \ldots, \mathbf{X}_n$ correspond to modes or peaks in a density estimate constructed from these points. The discussion below gives three rather different ways of putting this idea into practice, and is intended to be thought-provoking rather than prescriptive.

6.2.1 A hierarchical method

Several methods of cluster analysis work by arranging the objects in a tree structure of some kind. The tree may be of interest in its own right, as giving a hierarchical classification; alternatively, the data can be separated into individual clusters by breaking certain links in the tree and collecting together the objects in each connected component. Well-known clustering methods of this kind are the *single-linkage* and *complete-linkage* approaches, described, for example in Mardia, Kent and Bibby (1979, Chapter 11).

Density estimates can be used to define a hierarchical structure on a set of points in d-dimensional space. The objects will be arranged into a hierarchy by defining 'parent–child' relationships between them. The family tree, or trees, constructed from these relationships will give the hierarchical clustering of the data; at least one, and possibly more, of the objects will not have a parent.

Let $\hat{f}$ be a density estimate constructed from $\mathbf{X}_1, \ldots, \mathbf{X}_n$. Let d_{ij} be the Euclidean distance between $\mathbf{X}_i$ and $\mathbf{X}_j$. For each object $\mathbf{X}_i$, define a *threshold* t_i. Among objects within distance t_i of $\mathbf{X}_i$, choose as *parent* of $\mathbf{X}_i$ that object $\mathbf{X}_j$ which is steepest uphill from $\mathbf{X}_i$; in other words, choose j to maximize

$$\frac{\hat{f}(\mathbf{X}_l) - \hat{f}(\mathbf{X}_i)}{d_{il}} \tag{6.11}$$

over objects $\mathbf{X}_l$ for which

$$d_{il} \leqslant t_i \text{ and } \hat{f}(\mathbf{X}_l) > \hat{f}(\mathbf{X}_i). \tag{6.12}$$

If no points $\mathbf{X}_l$ satisfy (6.12) then $\mathbf{X}_i$ will have no parent and will be the root node of one of the family trees.

This algorithm is a slight generalization of the method suggested by Koontz, Narendra and Fukunaga (1976). Provided a suitable tie-breaking rule is used, it is easy to see that each $\mathbf{X}_i$ has at most one parent and that there can be no cycles in the hierarchy, in other words that $\mathbf{X}_i$ cannot be an ancestor of itself. If there are a large number of points $\mathbf{X}_i$, then a little thought shows that the effect of moving up the tree is to 'hill-climb' on the density estimate $\hat{f}$ towards a local maximum. Thus the divisions between families or clusters will tend to occur along 'valleys' in the density estimate.

The original authors give some encouraging examples together with some computational hints. They suggest two ways of constructing the density estimate $\hat{f}$ and the thresholds t_i. One suggestion is to use a kernel estimate for $\hat{f}$ and to set all the thresholds equal to the bandwidth h; the other is to use a kth-nearest neighbour estimate for $\hat{f}$ and to set the threshold t_i to be the distance from $\mathbf{X}_i$ to its kth nearest neighbour in the data set. In view of the spurious bumps likely to occur in the tails of fixed-width kernel density estimators, it is probably safest to use an adaptive kernel or nearest neighbour estimate for $\hat{f}$. There is no intrinsic reason why the choice of thresholds should be quite as closely tied to the construction of $\hat{f}$ as in the above suggestions. However, there is clearly some merit in linking the threshold values to the 'scale of pattern' to which $\hat{f}$ is sensitive. The extreme case, of essentially dispensing with the thresholds by setting $t_i = \infty$ for all i, may possibly cause difficulties by linking points in small clusters with strong modes a considerable distance away.

6.2.2 Moving objects uphill

A rather different approach to clustering is taken by Fukunaga and Hostetler (1975). In their approach, the objects themselves are allowed to move around in the space, and eventually become concentrated into a number of tight clumps. The algorithm works in an iterative way.

Suppose $\mathbf{X}_i^m$ is the position of object i at stage m of the procedure; initially $\mathbf{X}_i^0 = \mathbf{X}_i$. A density estimate $\hat{f}_{(m)}$ is constructed from the $\mathbf{X}_i^m$, and each point is moved in an uphill direction an amount proportional to the gradient of log $\hat{f}_{(m)}$ at that point. The constant of proportionality is a control parameter a that will be discussed below.

In symbols, we have

$$\mathbf{X}_i^{m+1} = \mathbf{X}_i^m + a\nabla \log \hat{f}_{(m)}(\mathbf{X}_i^m)$$

$$= \mathbf{X}_i^m + \frac{a\nabla \hat{f}_{(m)}(\mathbf{X}_i^m)}{\hat{f}_{(m)}(\mathbf{X}_i^m)} \tag{6.13}$$

by simple calculus. The quotient $\hat{f}_{(m)}(\mathbf{X}_i^m)$ in (6.13) does not affect the direction of the shift $\mathbf{X}_i^{m+1} - \mathbf{X}_i^m$, but it has several desirable properties. Points in the tails, where $\hat{f}_{(m)}$ is small, will be able to move a considerable distance towards regions of higher density. The step lengths will depend only on the shape of the local density; apart from errors involved in the density estimation, step lengths for points drawn from a particular cluster will depend only on the shape of the underlying distribution for that cluster, and not on the proportion of the total population for which it accounts. Furthermore, if $a = \sigma^2$ then points drawn from a spherical normal density with variance σ^2 will be moved to the centre of that distribution in a single step, again neglecting errors involved in the density estimation. Thus it is possible with a suitable choice of a for Gaussian clusters to be condensed into tight clumps very rapidly. Fukunaga and Hostetler (1975) give reasons why it is advisable to choose a conservatively; if a is chosen too large then objects are likely to 'overshoot' clusters and the algorithm may not converge.

In one very special case, the transformation (6.13) takes an interesting simple form. Suppose that the density estimates $\hat{f}_{(m)}$ are constructed by the kernel method using the multivariate Epanechnikov kernel (4.4) and that the control parameter a is set to the value

$$a = \frac{h^2}{d + 2}.$$

Then some simple algebra (given by Fukunaga and Hostetler, 1975) reduces (6.13) to the form

$$\mathbf{X}_i^{m+1} = \text{mean position of points } \mathbf{X}_j^m \text{ lying within Euclidean}$$
$$\text{distance } h \text{ of } \mathbf{X}_i^m \tag{6.14}$$

The derivation of (6.14) holds only if there are no points at exactly distance h from $\mathbf{X}_i^m$, but it can still be used as an algorithmic definition of $\mathbf{X}_i^{m+1}$ in this case. The point $\mathbf{X}_i^m$ itself is included in the set of points whose mean position is to be found. The algorithm given by (6.14) is called the *mean-shift algorithm* and has some attractive properties. A cluster of points of diameter less than h, which is separated from all other points by a distance of at least h, will be condensed to a common point at a single step. Furthermore, if the original data set is made up of convex subsets at least distance h apart, then these subsets will remain separated no matter how many iterations are performed.

An example of the application of the algorithm is given in Fig. 6.2. The original data were generated by taking 60 observations from each

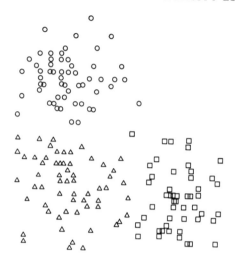

Fig. 6.2 *Clustering of a random data set by the mean-shift algorithm.*
Reproduced from Fukunaga and Hostetler (1975) with the permission of the
Institute of Electrical and Electronic Engineers, Inc. (Copyright © 1975
IEEE).

of three standard spherical normal distributions, centred at $(0, 0), (4, 0)$
and $(0, 4)$. Five applications of the mean-shift iteration (6.14) with
$h = 1.5$ led to the reduction of the data set to three distinct points
situated approximately at the means of the original distributions. The
corresponding assignment of the original data set is shown in the
figure. The algorithm can also be used in pattern recognition for
eliminating noise in images made up of points. An example is given in
Fig. 6.3 which shows a pattern of points before and after two
iterations of (6.13) with suitable parameters.

6.2.3 Sequence-based methods

It is often the case that the order in which the observations $\mathbf{X}_1, \ldots, \mathbf{X}_n$
are taken is an important one; for example, they might be obser-
vations taken sequentially in time. It may also happen that the
sequence violates the independence assumption in that neighbouring
observations in time tend to be close together in space, even though
the whole set $\{\mathbf{X}_1, \ldots, \mathbf{X}_n\}$ can be treated as an independent identically
distributed sample. In this event, a possible approach would be to

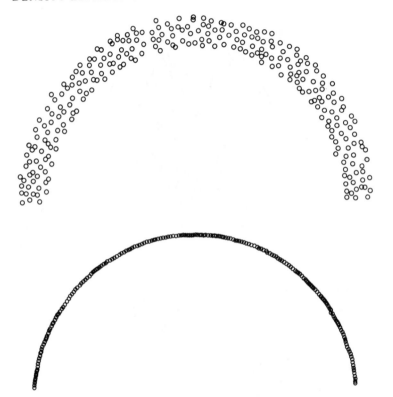

Fig. 6.3 *A data set before and after two mean-shift iterations. Reproduced from Fukunaga and Hostetler (1975) with the permission of the Institute of Electrical and Electronic Engineers, Inc. (Copyright © 1975 IEEE).*

consider the sequence $\hat{f}(\mathbf{X}_1)$, $\hat{f}(\mathbf{X}_2),\ldots,\hat{f}(\mathbf{X}_n)$ and to identify local maxima in this sequence as plausible cluster centres; the notion behind this idea is that the path traced out by the sequence in space is, roughly speaking, a continuous path that samples all 'important' parts of the density. Obviously, some selectivity and post-processing would be necessary, because the sequence $\{\hat{f}(\mathbf{X}_i)\}$ may contain a large number of local maxima, and a natural approach would be to feed the local maxima, together with their values of the density estimate, to an algorithm like the hierarchical approach described in Section 6.2.1.

Along similar lines to these is the approach of Kittler (1976). He

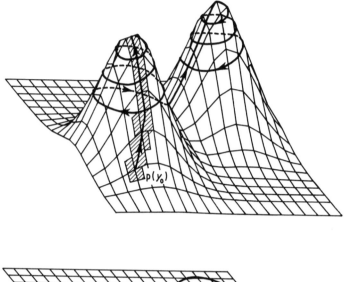

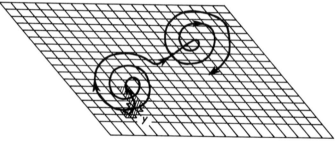

Fig. 6.4 *Path traced out on a density estimate by Kittler's approach.*
Reproduced from Kittler (1976) with the permission of Pergamon Press Ltd.

does not assume that the observations arise in any particular order,
but instead *arranges* the data in a sequence by reference to the values
of the density estimate. The rough aim is to construct a path through
the data set that takes in as many as possible of the data points near a
particular mode in $\hat{f}$ before going on to nearby modes. At each stage,
the next point is chosen from nearby data points in order to ensure
that one moves as far uphill (or as little downhill) on $\hat{f}$ as possible. This
gives a path that climbs steeply uphill to a mode and then descends
slowly, taking in a large number of nearby points. See Fig. 6.4 for an

illustration. Although perhaps a little idiosyncratic, the method is fast and appears to give reasonable results; see the original paper for more details.

6.3 Bump hunting and testing for multimodality

In Section 6.2, the interest in modes in the density estimate was that they indicated clustering in the data. No attempt was made to assess the statistical significance (in the widest sense) of these clusters; the approach was purely descriptive. Furthermore, the emphasis was on detecting clusters in sets of data rather than modes in underlying densities; though these two notions are somewhat indistinct, there is a slight difference of emphasis between them. It is a tautology to say that a mode in a density estimate corresponds to a cluster in the data from which the density estimate was constructed, since the concept of a 'cluster' is only defined by the method of cluster analysis being used. However, once the concept of a mode is defined, it is a clear statistical question whether an observed mode in a density estimate *really* arises from a corresponding feature in the assumed underlying density.

6.3.1 Definitions and motivation

In discussing the statistical assessment of observed bumps and modes, only the univariate case will be considered. Some of the remarks made have multivariate consequences, but these have not yet been investigated in any detail.

The definitions of bumps and modes in curves will be as follows: a *mode* in a density f will be a local maximum, while a *bump* will be an interval $[a, b]$ such that f is concave over $[a, b]$ but not over any larger interval. For smooth densities, this will correspond to an interval between a strict local maximum a and a strict local minimum b of the derivative f'.

The existence of more than one bump in a probability density has been discussed by Cox (1966), drawing on previous authors, as a 'descriptive feature likely to indicate mixing of components'. Of course it is meaningless merely to ask whether a density f is of the form $\pi_1 f_1 + \pi_2 f_2$ without specifying something about the likely component densities f_1 and f_2. Nevertheless, most standard densities do not have multiple bumps and so the presence of more than one bump in a density is indicative of a mixture. Furthermore, for the case of mixtures of normal components with equal variance, reasonable separation of the components will yield multiple bumps for a wide range of values of the mixing proportions π_1 and π_2; the minimum separation is approximately constant at 2 standard deviations for $0.2 \leqslant \pi_1 \leqslant 0.8$. Cox (1966) gives further discussion and references.

Good and Gaskins (1980) discuss a problem arising in high-energy physics where bumps in the underlying density of the observed data give evidence concerning elementary particles. They also remark that bumps 'indicate some feature of a random variable requiring an explanation'.

6.3.2 Two possible approaches based on density estimates

There have been several approaches suggested to the bump-hunting problem. Cox's (1966) proposal first requires a histogram to be constructed from the data; let f_i be the number of observations falling in cell i. The bin width of the histogram is to be chosen 'as broad as possible subject to the effects under study not being obscured'. If the true cell probabilities in cells $(i-1)$, i and $(i+1)$ are linearly related, then, conditionally on $f_{i-1} + f_i + f_{i+1}$, f_i has a binomial $(f_{i-1} + f_i + f_{i+1}, \frac{1}{3})$ distribution; hence the statistic t_i defined by

$$t_i = \frac{(f_{i+1} - 2f_i + f_{i-1})}{\sqrt{2(f_{i-1} + f_i + f_{i+1})}}$$

can be shown by the normal approximation to the binomial distribution to have approximately a standard normal distribution. (Cox, 1966, incorporates a continuity correction to yield a slightly different statistic.) Significant values of t_i indicate convexity or concavity of the true probability density function f around cell i, and examining the sequence of t_i's will yield some informal evidence about the number of bumps in f. The approach has the great advantage of being based on simple ideas, but has the disadvantages of the difficulty of interpretation of repeated significance tests, and the somewhat arbitrary choice of histogram bin width.

The approach of Good and Gaskins (1980) to the bump-hunting problem makes use of the Bayesian interpretation of the roughness penalty method as outlined in Section 5.4.5 above. In order to assess any particular bump in a maximum penalized likelihood estimate $\hat{f}$, a process called *iterative surgery* is used to construct another estimate

f^* in which the given bump is absent. The difference between the penalized log likelihood of $\hat{f}$ and that of f^* then gives the log odds factor in favour of the bump.

6.3.3 Critical smoothing

Silverman's (1981b) approach to the bump-hunting problem is quite different and is based on hypothesis testing. It has over Cox's (1966) proposal the advantages of not involving simultaneous significance tests and not requiring any prior specification of the scale of effects under consideration. The method involves the idea of 'critical smoothing' and is probably best explained in the context of a test for the unimodality of a density. Suppose, therefore, we have a sample $X_1, \ldots, X_n$ assumed to come from a density f and that we wish to test the hypothesis that f is unimodal against the hypothesis of multimodality. Consider the behaviour of the kernel estimate $\hat{f}$ with window width h, keeping the data points fixed but allowing the window width to vary. For very large h, one would expect $\hat{f}$ to be unimodal, while as h gets smaller there will come a point where the density becomes bimodal; decreasing h still further will allow more modes to appear while eventually there will be a very narrow kernel visibly centred at each data point. This behaviour can be described mathematically by saying that *the number of modes is a decreasing function of the window width h*. It turns out that such behaviour can be guaranteed only for certain kernel functions, and that the normal kernel is one of these; Silverman (1981b) gives a proof, which makes use of two very special properties of the normal distribution.

Thus, assuming that $\hat{f}$ is found by the kernel method using a normal kernel, there will indeed be a critical value h_{crit} of the smoothing parameter where the estimate changes from unimodal to multimodal. For all $h \geqslant h_{\text{crit}}$, $\hat{f}$ will be unimodal, while for all $h < h_{\text{crit}}$, $\hat{f}$ will be multimodal. Thus a simple binary search procedure can be used to find h_{crit} in practice; an interval $(h_{\text{lo}}, h_{\text{hi}})$ in which h_{crit} is known to fall can be halved in length by checking whether the value $\frac{1}{2}(h_{\text{lo}} + h_{\text{hi}})$ leads to a multimodal estimate. Thus h_{crit} can be found, to reasonable accuracy, by considering a fairly small number of estimates constructed from the given data. Step 5 of the algorithm described in Section 3.5 above makes it particularly easy to construct several estimates from the same data using a normal kernel.

For samples displaying clear bimodality, one might expect a large value of h_{crit}, because a considerable amount of smoothing will be needed to yield a unimodal density estimate. On the other hand, if the true underlying density f were unimodal, less smoothing would be required to give a unimodal estimate, and so a smaller value of h_{crit} would be obtained. Theoretical justification for these assertions is given by Silverman (1983), where it is proved, under mild conditions, that as the sample size n tends to infinity, h_{crit} will tend to zero if f is unimodal but will remain bounded away from zero otherwise. In summary, the critical value h_{crit} can be used as a statistic to investigate whether an unknown density is unimodal, and large values of h_{crit} will indicate multimodality.

Of course it remains to decide how large is 'large' in this context. One possibility is to assess h_{crit} against a standard family of unimodal densities. For instance, one could ask whether the observed value of h_{crit} would be unexpectedly large if the data came from a normal distribution with the same variance as the observed data. A simulation study reported by Jones (1983, Section 5.4) indicates that values of h_{crit} larger than $1.25\sigma n^{-1/5}$ will occur in only about 5% of samples drawn from normal distributions with variance σ. The choice (3.28) of window width $h = 1.06\sigma n^{-1/5}$ will give multimodal estimates about 15% of the time, while the rule (3.31) will give multiple modes in about one-third of the normal samples to which it is applied. These conclusions hold for a wide range of values of n (at least $40 \leqslant n \leqslant 5000$) and provide practical justification for theoretical results of Silverman (1983) on the asymptotic behaviour of h_{crit}.

If it is unattractive or clearly inappropriate to assess the observed value of h_{crit} against a standard family of distributions, then an alternative approach is possible; this is based on the ideas discussed in Section 6.4 and will be described there. The idea of critical smoothing can be extended in a natural way to provide a test for any given number k of modes (or bumps) against a larger number; the definition of h_{crit} becomes the smallest value of the window width yielding exactly k modes (or bumps) in the density estimate f. The observed value can be compared against the values obtained from standard densities by simulation, or alternatively the bootstrap approach discussed in Section 6.4 can be applied. See Silverman (1981b, 1983) for more details and an example.

There are broader contexts in which one can use, as a statistic, the most extreme value of a smoothing parameter at which a given feature occurs. It should be borne in mind that, in general, it may not be the case that the feature will be present for *all* less extreme parameter values; this can only be assumed with complete safety if, as in the discussion above, it can be proved in the particular context.

6.3.4 Discussion and further references

In this section, detailed attention has been restricted to methods that make direct use of density estimates. There are other important approaches to the problem of testing for multimodality. For example, Hartigan and Hartigan (1985) have proposed a technique called the DIP test where the statistic considered is the quantity D_H given by

$$D_H = \min_{F \text{ in } U} \sup_{x} |F_n(x) - F(x)|,$$

where F_n is the empirical (cumulative) distribution function of the data and U is the class of all distribution functions of unimodal densities. The general question of testing for clustering (as opposed to multimodality) is reviewed by Bock (1984, 1985), who provides numerous additional references.

It may be futile to expect very high power from procedures aimed at such broad hypotheses as unimodality and multimodality. Nevertheless, it is clear that much work remains to be done on investigating the detailed properties of the various methods discussed in this section, and on making comparisons between them.

6.4 Simulation and the bootstrap

The growing use of computer-intensive simulation approaches to problems in statistics and operational research leads to an important question that has been given surprisingly little attention in the literature. Suppose we want to simulate from a distribution but that all the information we have about the distribution is a set of observations drawn from it. A natural context for this problem is the modelling of complex systems in operational research, where the information on individual components of the systems may well be obtained by observing their behaviour over a period, and so consists of samples from distributions rather than exact knowledge of the distributions involved. A more purely statistical context is the bootstrap methodology discussed further in Section 6.4.2. First of all, some general remarks about the use of density estimates in simulation will be made. The availability of pseudo-random numbers from various standard distributions will be assumed; for details of how these are generated in practice, see, for example, Ripley (1983) and the references cited there.

6.4.1 *Simulating from density estimates*

An idealized form of the problem under consideration in this section is easily formulated mathematically. Given a sample $\mathbf{X}_1, \ldots, \mathbf{X}_n$ from an unknown density f, we are required to construct a sequence of independent observations $\mathbf{Y}_1, \mathbf{Y}_2, \ldots$ from f. Of course, this will be impossible to achieve exactly in practice because full information about f is not available. The observations in the sample $\{\mathbf{X}_i\}$ and the required realizations $\{\mathbf{Y}_j\}$ will be assumed to be d-dimensional vectors.

The usual approach to the problem is one of two extremes. A parametric form for f could be assumed, such as the normal distribution with unknown parameters; the sample $\{\mathbf{X}_i\}$ is used to estimate the unknown parameters, and a standard simulation method is then used to generate the required simulated observations. Alternatively, the realizations $\{\mathbf{Y}_j\}$ are generated directly by successive random sampling, with replacement, from the sample $\{\mathbf{X}_1, \ldots, \mathbf{X}_n\}$. This latter approach has the advantage of freeing the procedure from parametric assumptions but the serious disadvantage, in some contexts, of making it impossible for any value to occur in the simulated data that has not occurred exactly in the original sample $\{\mathbf{X}_i\}$. This behaviour may be unacceptable because spurious very fine structure in the original data may well be faithfully reproduced in the simulated samples. This may not present a problem in practice, depending on the context. However, difficulties will arise if the ideal aim of the simulation is to produce samples that have the underlying 'true' structure of the observed data without sharing spurious details that have arisen from random effects.

It is very natural to consider an intermediate approach based on density estimation. The observations $\mathbf{X}_1, \ldots, \mathbf{X}_n$ can be used to construct a nonparametric estimate $\hat{f}$ of the density f, and then as many independent realizations as required can be drawn from $\hat{f}$. Depending on the context, it may well be desirable to simulate not from $\hat{f}$ itself but from a version transformed to have the same mean vector and covariance matrix as the observed data; we shall return to this refinement below.

If $\hat{f}$ is constructed by the kernel method or the adaptive kernel method, then it is very easy to find independent realizations from $\hat{f}$, provided a non-negative kernel is used. Indeed, it is not even necessary to find $\hat{f}$ explicitly in the simulation procedure. Consider the univariate case, and suppose $\hat{f}$ has been constructed by the ordinary kernel method with kernel K and window width h. Realizations Y from $\hat{f}$ can be generated as follows:

Step 1 Choose I uniformly with replacement from $\{1, \ldots, n\}$
Step 2 Generate ε to have probability density function K $\left.\right\}$ (6.15)
Step 3 Set $Y = X_I + h\varepsilon$.

It is necessary in Step 2 to generate a random observation from the kernel K. If K is the normal density, there are several standard ways of doing this, such as the Box–Muller and Marsaglia approaches discussed, for example, in Ripley (1983, Section 4.2). Devroye and Györfi (1985, p. 236) give a very fast algorithm for simulation from the rescaled Epanechnikov kernel

$$K(x) = \tfrac{3}{4}(1 - x^2) \text{ for } |x| \leqslant 1$$

as follows:

Step 2a Generate three uniform $[-1, 1]$ random variates V_1, V_2, V_3.

If $|V_3| \geqslant |V_2|$ and $|V_3| \geqslant |V_1|$, set $\varepsilon = V_2$;

otherwise set $\varepsilon = V_3$.

The algorithm (6.15) can be repeated as often as necessary to give independent realizations Y_j from $\hat{f}$. It is fast and easy to program. Proofs that (6.15) really does give an observation from $\hat{f}$, and that Step 2a has the desired effect, are exercises in elementary probability theory and are omitted. A multivariate version of (6.15) is easily constructed; unequal smoothing parameters in the various coordinate directions, or the use of a matrix of shrinking coefficients, are easily coped with by using the corresponding transformation in Step 3.

If the realizations Y are required to have first and second moment properties the same as those observed in the sample $\{X_1, \ldots, X_n\}$, then Step 3 in (6.15) should be replaced by

Step 3' $Y = \bar{X} + (X - \bar{X} + h\varepsilon)/(1 + h^2\sigma_K^2/\sigma_X^2)^{1/2}$ (6.16)

where $\bar{X}$ and σ_X^2 are the sample mean and variance of $\{X_i\}$ and σ_K^2 is the variance of the kernel K. In the multivariate case, the step corresponding to Step 3' is simplest if the kernel is scaled to have variance matrix the same as the data; it then becomes

$$Step\ 3''\quad \mathbf{Y} = \bar{\mathbf{X}} + (\mathbf{X} - \bar{\mathbf{X}} + h\boldsymbol{\varepsilon})/(1 + h^2)^{1/2}. \tag{6.17}$$

Notice that the original algorithm (6.15) will yield realizations with expected value equal to $\bar{X}$, but that the smoothing will increase the variance unless a correction like Step 3' or 3'' is used.

Modifying (6.15) to give realizations from an adaptive kernel estimate is straightforward. The last step is replaced by

$$Step\ 3^*\quad Set\ Y = X_I + h\lambda_I\varepsilon \tag{6.18}$$

where h and $\lambda_1, \ldots, \lambda_n$ are defined as in (5.7) and (5.8) above.

Some further discussion of the material of this section is given in Devroye and Györfi (1985, Section 8). They give some theoretical background that shows that, perhaps not surprisingly, very large sample sizes n are required if one is to be confident that moderately long sequences $Y_1, \ldots, Y_m$ are to be practically indistinguishable, in all respects, from sequences generated from the true density f. They also give some additional algorithms for simulation from density estimates of various kinds.

6.4.2 The bootstrap and the smoothed bootstrap

The bootstrap is an appealing approach to the assessment of errors and related quantities in statistical estimation. The method is described and explored in detail by Efron (1982), and only a brief explanation will be given here.

Let $\rho(F)$ be some interesting property of a distribution F that depends in some complicated way on F. Typically, even if F is known, ρ can most easily be estimated by repeatedly simulating samples from F. For example, ρ might be the sampling variance of the upper quartile of samples of size 39 drawn from F; a way of finding ρ to within reasonable accuracy is to simulate 1000 samples of size 39 from F, to find the upper quartile of each of these samples, and then to calculate the sample variance of these 1000 values.

In many statistical problems, F itself is unknown but a sample $X_1, \ldots, X_n$ of observations from F is available. The standard *bootstrap* approach is to estimate $\rho(F)$ using the procedure just described, but to simulate the samples not from F itself but from the empirical distribution function F_n of the observed data $X_1, \ldots, X_n$. A sample from F_n is generated by successively selecting uniformly with replacement from $\{X_1, \ldots, X_n\}$. This approach approximates $\rho(F)$ by $\rho(F_n)$, and an example will be given below.

The samples constructed from F_n in the bootstrap simulations will have some rather peculiar properties. All the values taken by members of these samples will be drawn from the original values $X_1, \ldots, X_n$ and nearly every sample will contain repeated values. If n is at all large, most samples will contain some values repeated several times.

An approach that does not lead to samples with these properties is the *smoothed bootstrap*. Here the simulations are constructed not from F_n but by using an algorithm like (6.15) to simulate from a smoothed version of F_n. If $\hat{F}$ is the distribution function of the density estimate $\hat{f}$, then the effect of the smoothed bootstrap will be to estimate $\rho(F)$ by $\rho(\hat{F})$. Whether $\rho(F)$ is better estimated by $\rho(F_n)$ or $\rho(\hat{F})$ will depend on the context

An example where the smoothed bootstrap does well is given by Efron (1981). Let F be a bivariate distribution of random vectors $\mathbf{X} = (\xi, \eta)$, and let $\phi(F)$ be Fisher's variance-stabilized transformed correlation coefficient

$$\phi(F) = \tanh^{-1} \mathrm{corr}(\xi, \eta). \qquad (6.19)$$

Given a sample $\mathbf{X}_1, \ldots, \mathbf{X}_{14}$, an estimate $\hat{\phi}(\mathbf{X}_1, \ldots, \mathbf{X}_{14})$ is constructed by substituting the usual sample correlation coefficient into (6.19). Let $\rho(F)$ be the sampling standard deviation of $\hat{\phi}(\mathbf{X}_1, \ldots, \mathbf{X}_{14})$ if $\mathbf{X}_1, \ldots, \mathbf{X}_{14}$ are drawn from F. Efron (1981) considered the case where F is a bivariate normal distribution with $\mathrm{var}(\xi) = \mathrm{var}(\eta) = 1$ and $\mathrm{cov}(\xi, \eta) = \frac{1}{2}$. For this case, the true value of $\rho(F)$ is 0.299. In 200 Monte Carlo trials, the bootstrap estimated $\rho(F)$ with root-mean-square error 0.065, while the smoothed bootstrap had a root-mean-square error of only 0.041, a substantial improvement. The smoothed bootstrap was implemented using the algorithm (6.15) with a normal kernel with the same covariance matrix as the data; Step $3''$ of (6.17) was used to ensure that $\hat{F}$ had the same second-order characteristics as F_n. The smoothing parameter h in (6.17) was set to 0.5.

There has, as yet, been very little systematic investigation of the circumstances under which the smoothed bootstrap will give better results than the ordinary bootstrap. In addition, the choice of smoothing parameter in the smoothed bootstrap is usually made completely arbitrarily; detailed work on this choice, making use of known properties of density estimates, would no doubt improve the performance of the smoothed bootstrap. A perhaps somewhat controversial problem where the smoothing parameter is chosen in a natural way is discussed in the next section.

6.4.3 A smoothed bootstrap test for multimodality

The smoothed bootstrap can be used to construct a test for multimodality based on the critical smoothing idea discussed in Section 6.3.3 above. Given a data set $X_1,\ldots,X_n$ yielding a value h_0 for the critical window width h_{crit}, one needs to decide whether h_0 is a surprisingly large value for the statistic h_{crit}. To do this, h_0 has to be assessed against some suitable unimodal null density f_0 for the data. An approach suggested in Section 6.3.3 was to use a member of a standard parametric family for f_0, but the smoothed bootstrap makes a more nonparametric approach possible.

A suitable choice of f_0 would have various desirable properties, as follows:

(a) The density f_0 must be unimodal, since f_0 must be a representative of the compound null hypothesis of unimodality.
(b) Subject to (a), f_0 should be a plausible density underlying the data; testing against all possible unimodal densities is a hopeless task, since, for example, large values of h_{crit} would be obtained from unimodal densities with very large variances.
(c) In order to give unimodality a fair chance of explaining the data, f_0 should be, in some sense, the most nearly bimodal among those densities satisfying (a) and (b).

A very natural way of constructing f_0 to satisfy these requirements is to set f_0 equal to the density estimate $\hat{f}_{crit}$ constructed from the original data with window width h_0. It can be shown (see Silverman, 1981b) that $\hat{f}_{crit}$ is unimodal; as a density estimate constructed from the data it is a plausible density for the data; and, by the definition of h_{crit}, reducing the window width any further would make $\hat{f}$ multimodal, so $\hat{f}_{crit}$ is an 'extreme' unimodal density. In fact it can be shown that, as the sample size increases, the asymptotic behaviour of h_{crit} is such as to yield good estimates of the true density f if f is indeed unimodal; see Silverman (1983).

Sampling from $\hat{f}_{crit}$ corresponds precisely to the smoothed bootstrap algorithm (6.15) with h equal to h_0. In assessing the significance value of h_0, it is necessary only to ascertain the proportion of samples of size n from $\hat{f}_{\text{crit}}$ that lead to values of h_{crit} greater than h_0. A given sample will have $h_{\text{crit}} > h_0$ if and only if the density estimate constructed from the sample with window width h_0 is multimodal. Thus an algorithm for assessing the p-value of h_0 is to generate a large number of samples from $\hat{f}_{\text{crit}}$ and to count the proportion of samples which yield a multimodal density estimate using window width h_0. There is no need to find h_{crit} for each sample.

An example of the application of this technique is given in Silverman (1981b). A sample of 22 observations on chondrite meteors was investigated for multimodality. The original data are given in Good and Gaskins (1980, Table 2). The estimated significance value of h_{crit} for unimodality was 8%, calculated by the procedure discussed in this section. For further details the reader is referred to Silverman (1981b), but warned that the p-values printed in Table 1 of that paper are incorrect and should all be subtracted from 1.

6.5 Estimating quantities that depend on the density

Often it is not so much the density itself that is of interest but some curve or single numerical quantity that can be derived from the density. An important example of a curve that depends on the density is the hazard rate, the estimation of which will be discussed in Section 6.5.1. A single numerical quantity that depends on the density is called a *functional* of the density; a general discussion of the estimation of functionals of the density will be given in Section 6.5.2, and a particular application will be developed in Section 6.5.3.

6.5.1 *Hazard rates*

Given a distribution with probability density function f, the *hazard rate* $z(t)$ is defined by

$$z(t) = \frac{f(t)}{\int_t^\infty f(u)\mathrm{d}u} \tag{6.20}$$

The hazard rate is also called the age-specific or conditional failure rate. It is useful particularly in the contexts of reliability theory and survival analysis and hence in fields as diverse as engineering and medical statistics. Cox and Oakes (1984), for example, give a discussion of the role of the hazard rate in understanding and modelling survival data. They also provide a survey of some methods (other than those discussed here) for the nonparametric estimation of hazard rates.

The formula (6.20) can be rewritten

$$z(t) = \frac{f(t)}{1 - F(t)} \qquad (6.21)$$

where $F(t)$ is the cumulative distribution function. Given a sample $X_1, \ldots, X_n$, a very natural approach to the nonparametric estimation of the hazard rate is to construct the estimate

$$\hat{z}(t) = \frac{\hat{f}(t)}{1 - \hat{F}(t)} \qquad (6.22)$$

where $\hat{f}$ is a suitable density estimate based on the data and

$$F(t) = \int_{-\infty}^{t} \hat{f}(u)\,du.$$

If $\hat{f}$ is constructed by the kernel method, then $\hat{F}(t)$ can be written

$$\hat{F}(t) = n^{-1} \sum_{i=1}^{n} \mathscr{K}\{(t - X_i)/h\}$$

where $\mathscr{K}(u)$ is the cumulative distribution function of the kernel,

$$\mathscr{K}(u) = \int_{-\infty}^{u} K(v)\,dv.$$

For most kernels (except the Gaussian) $\mathscr{K}$, and hence the estimate (6.22), can be found explicitly. The estimate (6.22) can also be found very easily if $\hat{f}$ and hence $\hat{F}$ are estimated by the adaptive kernel method of Section 5.3.

Watson and Leadbetter (1964) introduced and discussed $\hat{z}(t)$ and various alternative nonparametric estimators of $z(t)$. They gave the asymptotic formula

$$\text{var}\{\hat{z}(t)\} \approx \frac{1}{nh} \frac{z(t)}{\{1 - F(t)\}} \int K(v)^2\,dv \qquad (6.23)$$

$$= \frac{1}{nh} \frac{z(t)^2}{f(t)} \int K(v)^2\,dv \qquad (6.24)$$

valid, under suitable conditions, when f is estimated by the kernel method with kernel K and window width h; simulations reported by Rice (1975) show that (6.24) gives a good approximation even for sample sizes that are relatively small in this context, $n = 100$ or more. For further properties of $\hat{z}(t)$ and related estimators, see, for example, Rice (1975), Rice and Rosenblatt (1976) and Prakasa Rao (1983, Section 4.3). An interesting direct approach based on penalized likelihood is given by Anderson and Senthilselvan (1980).

Consideration of the errors involved in the construction of $\hat{z}$ show that, to a first approximation, the main contribution to the error will be due to the numerator in (6.22). Thus to get the best possible estimate of the hazard, one should aim to minimize the error in the estimation of the density. However, it must be pointed out that (6.22) and the form of (6.23) show that the probable error of the hazard estimate will blow up in the right-hand tail of the distribution, and that relatively small fluctuations in $\hat{f}(t)$, for large t, will become much larger when divided by the quotient $1 - \hat{F}(t)$. This magnification of errors in the tail, combined with the fact that long-tailed distributions are the rule rather than the exception in survival and reliability data, makes it advisable to use a method like the adaptive kernel method rather than the fixed-width kernel method in this context. If an adaptive method of the form described in Section 5.3.1 is used then the approximate variance (6.24) will become

$$\text{var } \hat{z}(t) \approx \frac{1}{nhg^{\alpha}} \frac{z(t)^2}{f(t)^{1-\alpha}} \int K(v)^2 \mathrm{d}v, \qquad (6.25)$$

which will increase less rapidly as t increases; furthermore, using the value $\alpha = \frac{1}{2}$ will, as noted in Section 5.3.2, have the advantage of reducing the bias generally.

An example of a hazard rate estimate is given in Fig. 6.5, constructed from the suicide data of Table 2.1. The hazard rate estimate has been found from the adaptive kernel estimate shown in Fig. 5.3; the lower curve gives an estimate of the (pointwise) sampling standard deviation of the estimated hazard, obtained by substituting the estimates of z and f back into (6.25). Consideration of both curves together suggests that there is a genuine fall in hazard between 100 and 500, but that the subsequent increase may be a purely random effect. Including the standard deviation curve is useful in this context because it stresses the unreliability of the estimates for large values of t. The curves corresponding to Fig. 6.5 constructed

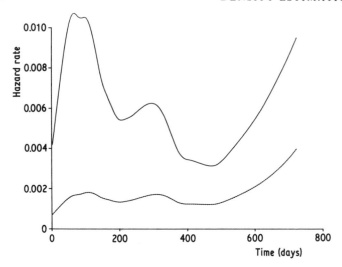

Fig. 6.5 *Hazard rate estimate (upper curve) and estimated sampling standard deviation of the estimate (lower curve) constructed from the suicide data, based on the adaptive kernel density estimate shown in Fig. 5.3.*

from a fixed-width kernel estimate display the same general shape, but with much wider fluctuation from 400 upwards and much more rapid increase in standard deviation in that range.

A variant of the estimator (6.22) is given by using the empirical distribution function $F_n(t)$ of the data in the denominator instead of the smoothed version $\hat{F}(t)$. Because of the jumps in $F_n(t)$, this estimator contains corresponding local variations. The two estimators have the same asymptotic bias and variance, and so the form (6.22) is probably preferable because it yields smoother curves free of unnecessary local fluctuations. An example constructed using the alternative estimator is given in Fig. 6.6, reproduced from Rice (1975). The underlying data consist of 4763 intervals of time between successive micro-earthquakes in an area in California. A fixed-width kernel density estimate constructed with a triangular kernel was used, but a much larger window width was used for the range 60–1560 than for the range 0–100. The small-scale fluctuations can be shown, by constructing standard deviations as in Fig. 6.5, to be merely noise; however, it is clear that the hazard drops quickly up to time 10 and then continues to fall gently. A further curve given by Rice (1975) for range 1600–3600 indicates that after about time 900 the hazard

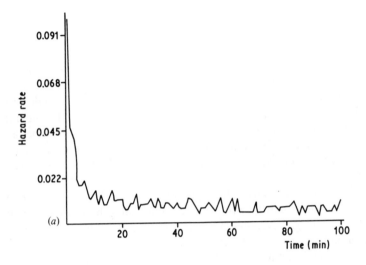

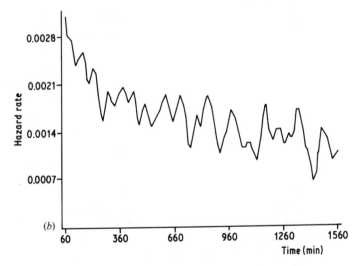

Fig. 6.6 *Hazard rate estimates constructed from 4763 successive intervals of time between micro-earthquakes. Reproduced from Rice (1975) with the permission of the Royal Astronomical Society and the author.*

appears to have levelled off. Rice (1975) goes on to explain how these curves suggest a Poisson cluster model for the original earthquake sequence.

6.5.2 Functionals of the density

There are a number of statistical problems where it is necessary to obtain estimates of functionals like $\int f^2$ or $\int f \log f$. An obvious way of constructing these estimates is to find an estimate $\hat{f}$ of f and then to substitute this estimate into the required integral. Usually the integral will have to be evaluated by numerical quadrature, but sometimes an exact formula can be derived. For example, if $\hat{f}$ is a kernel estimate then $\int \hat{f}^2$ can be expressed explicitly using the formula (3.37) given above.

The functional $\int f(t)^{1/4} dt$ is estimated using a density estimate in work of Silverman (1984b) on the spline smoothing approach to nonparametric regression mentioned in Section 5.4.5. Given a set of data pairs (t_i, Y_i), the smoothing parameter α in (5.33) can be chosen automatically by a cross-validation method. An approximation to the cross-validation score can be obtained involving $\int f(t)^{1/4} dt$, where $f(t)$ is the local density of the design set $\{t_1, \ldots, t_n\}$ at t. Using a kernel density estimate for f makes it possible to calculate the approximation to the cross-validation score extremely quickly, and hence the automatic choice of the smoothing parameter can be speeded up enormously. In addition it turns out that the statistical performance of the approximate cross-validation score is actually better than that of the exact score. For full details and practical examples see Silverman (1984b, 1985a).

The smoothed bootstrap discussed in Section 6.4.2. can be viewed as a rather complicated example of the estimation of a functional of a density. The quantity ρ of interest could, in most cases, only be discerned from $\hat{f}$ by a simulation procedure, since the dependence of ρ on $\hat{f}$ could not be expressed in a simple, explicit way.

The final section of this chapter describes an exciting application of density estimation ideas in data analysis. There are no doubt many other contexts in which the use of density estimates to estimate functionals of the density would give useful results, and it is hoped that this section will provide some impetus for the development of further applications. In general, for the estimation of functionals of the density, the choice of smoothing parameter is (perhaps not surprisingly) not as sensitive as for the estimation of the density itself. However, the need for an automatic method of choice is often more pressing. In the author's experience, approaches based on reference to standard distributions, as discussed in Section 3.4.2, usually give good results if applied carefully. More sophisticated methods like cross-validation are rarely warranted in this context.

6.5.3 Projection pursuit

Suppose we are given a cloud of data points observed in a fairly high-dimensional space. For many purposes, not least for exploration and presentation of the data, it is useful to reduce the dimensionality of the data by projecting them linearly onto a one- or two-dimensional subspace. A classical technique for dimension reduction of this kind arises in *principal components analysis*; reduction of the data to one or two principal components corresponds precisely to projection onto a subspace chosen to maximize the variance(s) of the projected data. Detailed descriptions of principal components analysis are given by Chatfield and Collins (1980) and Mardia, Kent and Bibby (1979).

A more general strategy for dimension reduction is to define what is meant by an 'interesting' configuration of the data set and then to select the subspace onto which the projection of the data is most interesting. Principal components analysis can be viewed in this framework by deeming interesting configurations to be those with high variances. For many problems, principal components analysis has given excellent results; however, it is obvious that high variance is not necessarily the only relevant criterion of the genuine importance of structure displayed in a projection of the data. The general idea of *projection pursuit* is to define some other criterion of the interest of a projected configuration and then to use a numerical optimization technique to find the projection of most interest.

An important early paper on projection pursuit is by Friedman and Tukey (1974), who coined the term projection pursuit and provided many valuable insights on which subsequent work has been based; a clear digest of the Friedman–Tukey method is given in Tukey and Tukey (1981, Section 11.2.4). Significant recent contributions include Jones (1983), Jones and Sibson (1987) and Huber (1985).

Suppose that the original d-dimensional data set is $\{\mathbf{X}_1, \ldots, \mathbf{X}_n\}$. A one-dimensional projection is defined by selecting a unit d-vector $\mathbf{p}$ and finding $\mathbf{p}^T\mathbf{X}_1, \ldots, \mathbf{p}^T\mathbf{X}_n$. A two-dimensional projection is defined by selecting a $(d \times 2)$ matrix P, the columns of which are orthogonal unit d-vectors, and finding $P^T\mathbf{X}_1, \ldots, P^T\mathbf{X}_n$. In this section, the projected data set will be written as $\xi_1, \ldots, \xi_n$ where the ξ_i are points in one- or two-dimensional space. Of course the ξ_i depend on the vector $\mathbf{p}$ or the matrix P, but this dependence will not be expressed explicitly.

The merit of a particular configuration $\{\xi_1, \ldots, \xi_n\}$ in exposing important structure in the data will be quantified by means of a *projection index* $I(\xi_1, \ldots, \xi_n)$. Since the points ξ_i depend on the projection, we shall sometimes abuse notation and write $I(\mathbf{p})$ or $I(P)$ as appropriate. Principal components reduction effectively uses sample variance as a projection index. Most generally, if we have in mind an index $q(f)$ of the interest of a given density f, then a projection index can be formulated by setting

$$I(\xi_1, \ldots, \xi_n) = q(\hat{f}) \qquad (6.26)$$

where $\hat{f}$ is a density estimate constructed from $\xi_1, \ldots, \xi_n$. Most, if not all, of the projection indices in the literature can be viewed in this way, though this was not necessarily the original motivation. A philosophical advantage of using a projection index of the form (6.26) is that projection pursuit based on a set of data can be thought of as a 'sample version' of projection pursuit based on a multivariate probability distribution. Given a multivariate distribution F, let $f_\mathbf{p}$ be the marginal density of F projected on the direction $\mathbf{p}$; maximizing $q(f_\mathbf{p})$ over all unit vectors $\mathbf{p}$ is the 'population version' of projection pursuit based on (6.26).

Friedman and Tukey (1974) used an index that is sensitive both to the spread of the projected sample and to clustering within it. However, more recent suggestions have concentrated on information separate from that presented by principal components, either by pre-processing the original data to have unit covariance matrix, or by concentrating on functionals $q(f)$ that are invariant under scale transformations of f. In our subsequent discussion we shall assume either that the data have been pre-processed or that f has variance 1 and that, if necessary, f has been scale-transformed to achieve this. Concentrate on the case of one-dimensional projections for the moment.

Two attractive projection indices are the negative Shannon entropy

$$q_1(f) = \int f \log f \qquad (6.27)$$

and the functional

$$q_2(f) = \int f^2. \qquad (6.28)$$

Among densities with unit variance, q_1 is minimized by the normal density, and so gives a measure of non-normality of f. The functional q_2 is minimized, as was pointed out in Section 3.3.2, by setting f equal to the Epanechnikov kernel function (3.24); thus basing a projection index on q_2 will yield projections whose underlying densities are 'least parabolic' rather than 'least normal'. Friedman and Tukey (1974) essentially used the functional q_2, while in a study of practical examples Jones (1983) found little difference between the results obtained using q_1 and q_2.

The maximization of the projection index $I(\mathbf{p})$ is facilitated enormously if an appropriate density estimate $\hat{f}$ is used in (6.26). The index is a function of a unit d-vector, and thus the pursuit algorithm involves a constrained numerical optimization in d-dimensional space. To carry out projection pursuit onto a two-dimensional subspace, the $(d \times 2)$ matrix P has to be found, and so a $2d$-dimensional optimization is involved. These optimizations are much easier to perform numerically by hill-climbing techniques if the index I is a smooth function of $\mathbf{p}$ (or P) and if the derivatives of I can be expressed explicitly. Since the projected points ξ_i are linear functions of the vector $\mathbf{p}$, the essential property is that the density estimate $\hat{f}$ depends smoothly on the underlying data points from which it is constructed.

Suppose, for example, that $\hat{f}$ is a kernel estimate based on the Gaussian kernel and $q(f) = \int f \log f$. Then I will be a smooth function of the projection direction $\mathbf{p}$ and we will have, by elementary calculus,

$$\frac{\partial I}{\partial p_r} = \int \{1 + \log \hat{f}(x)\} \frac{\partial \hat{f}(x)}{\partial p_r} \, dx$$
$$= \int \{1 + \log \hat{f}(x)\} \sum_i X_{ri} n^{-1} h^{-2} \phi'\{h^{-1}(\xi_i - x)\} \, dx \qquad (6.29)$$

where X_{ri} is the rth component of the ith data point $\mathbf{X}_i$. The integrand in (6.29) can be calculated quickly using extensions of the Fourier transform technique described in Section 3.5.

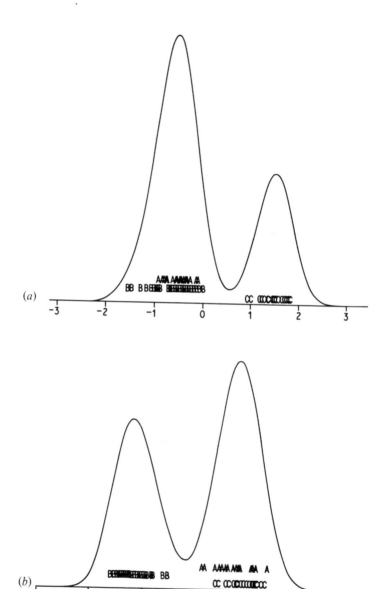

Fig. 6.7 *Projected beetle data and density estimates in two important directions. (a) Global maximum of projection index. (b) Next strongest local maximum. Reproduced from Jones (1983) with the permission of the author.*

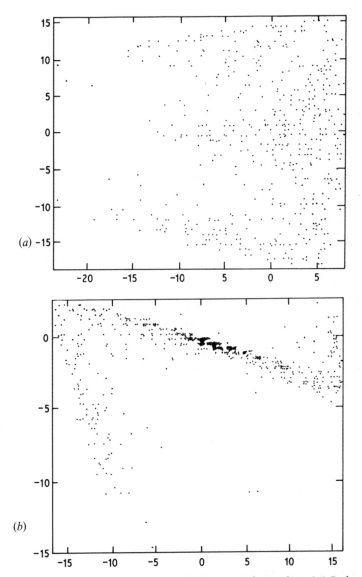

Fig. 6.8 *Two-dimensional projections of high-energy physics data. (a) Reduction to two leading principal components. (b) Global maximum of projection index. Reproduced from Friedman and Tukey (1974) with the permission of the Institute of Electrical and Electronic Engineers, Inc. (Copyright © 1974 IEEE).*

Insights from the general theory of density estimation can be used to choose a smoothing parameter h in the construction of $\hat{f}$. In an empirical study involving real and simulated data, Jones (1983) found that, if the data are pre-processed to have unit variance and a normal kernel is used, then good results were obtained using the value $h = n^{-1/5}$, suggested by the discussion of Section 3.4.2 above. The broad relative behaviour of the index $I(P)$ was not enormously sensitive to the choice of h, except that small values of h could give rise to numerical difficulties caused by spurious local maxima in the projection index. However, clear local maxima of the index (present at larger values of h) often correspond to projections displaying important structure.

To conclude this discussion, two examples are given which illustrate the power of the technique. Lubischew (1962, Tables 4, 5 and 6) gives six measurements on each of 74 beetles; these fall into three species which we will call A, B and C. Jones (1983) investigated the effect of first 'sphering' the data set by transforming it to have unit variance matrix and then applying the projection pursuit algorithm. Fig. 6.7, reproduced from Jones (1983, Figs 6.8 and 6.10), shows density estimates of the projected data in the directions corresponding to the global maximum and the next strongest local maximum of the projection index. The species of the beetles corresponding to the individual projected data points are also shown on the figure; of course the species information was not used in the projection pursuit process. It can be seen that the two projections together can be used to divide the data perfectly into the three species. Projections of the original data in the first two principal component directions give far inferior information about clustering in the data. In this example, the projection index was as in (6.26), using the functional q_1 of (6.27) and a normal kernel density estimate with window width $h = 74^{-1/5}$.

The second example is reproduced from Friedman and Tukey (1974) and illustrates projection pursuit onto a two-dimensional subspace. The underlying data consist of 500 seven-dimensional observations taken in a particle physics experiment. Full details of the data set and of the projection pursuit algorithm are given in the original paper. The scatter plots in Fig. 6.8 show projections of the data onto the first two principal components and onto a plane found by projection pursuit. The projection pursuit view shows some interesting structure that is invisible in the principal components view.

Bibliography

Abramson, I.S. (1982). On bandwidth variation in kernel estimates – a square root law. *Ann. Statist.*, **10**, 1217–1223.

Adler, R.J. and Firman, D. (1981). A non-Gaussian model for random surfaces. *Phil. Trans. Roy. Soc. Lond. A*, **303**, 433–462.

Aitchison, J. and Aitken, C.G.G. (1976). Multivariate binary discrimination by the kernel method. *Biometrika*, **63**, 413–420.

Anderson, J.A., Whaley, K., Williamson, J. and Buchanan, W.W. (1972). A statistical aid to the diagnosis of Keratoconjunctivitis sicca. *Quart. J. Med.*, **41**, 175–189.

Anderson, J.A. and Senthilselvan, A. (1980). Smooth estimates for the hazard function. *J. Roy. Statist. Soc. B.*, **42**, 322–327.

Bartlett, M.S. (1963). Statistical estimation of density functions. *Sankhya Ser. A*, **25**, 245–254.

Bean, S.J. and Tsokos, C.P. (1980). Developments in nonparametric density estimation. *Int. Stat. Rev.*, **48**, 267–287.

Bertrand-Retali, M. (1978). Convergence uniforme d'un estimateur de la densité par la méthode de noyau. *Rev. Roumaine Math. Pures. Appl.*, **23**, 361–385.

Bickel, P.J. and Rosenblatt, M. (1973). On some global measures of the deviation of density function estimates. *Ann. Statist.*, **1**, 1071–1095.

Bock, H.H. (1984). Statistical testing and evaluation methods in cluster analysis. *Statistics : Applications and New Directions. Proceedings of the Indian Statistical Institute Golden Jubilee International Conference*, 116–146.

Bock, H.H. (1985). On some significance tests in cluster analysis. *J. Classification*, **2**, 77–108.

Boneva, L.I., Kendall, D.G. and Stefanov, I. (1971). Spline transformations: three new diagnostic aids for the statistical data-analyst (with Discussion). *J. Roy. Statist. Soc. B*, **33**, 1–70.

Bowman, A.W. (1984). An alternative method of cross-validation for the smoothing of density estimates. *Biometrika*, **71**, 353–360.

Bowman, A.W. (1985). A comparative study of some kernel-based nonparametric density estimators. *J. Statist. Comput. Simul.*, **21**, 313–327.

Bowman, A.W., Hall, P. and Titterington, D.M. (1984). Cross-validation in nonparametric estimation of probabilities and probability densities. *Biometrika*, **71**, 341–351.

Bowyer, A. (1980). Experiments and computer modelling in stick-slip. Ph.D. Thesis, University of London.

Breiman, L., Meisel W. and Purcell, E. (1977). Variable kernel estimates of multivariate densities. *Technometrics*, **19**, 135–144.

Breiman, L., Friedman, J.H., Olshen, R.A. and Stone, C.J. (1984). *Classification and Regression Trees*. Belmont, California: Wadsworth.

Cacoullos, T. (1966). Estimation of a multivariate density. *Ann. Inst. Statist. Math.*, **18**, 179–189.

Čencov, N.N. (1962). Evaluation of an unknown distribution density from observations. *Soviet Math.*, **3**, 1559–1562.

Chatfield, C. and Collins, A.J. (1980). *Introduction to Multivariate Analysis*. London: Chapman and Hall.

Chow, Y.S., Geman, S. and Wu, L.D. (1983). Consistent cross-validated density estimation. *Ann. Statist.*, **11**, 25–38.

Copas, J.B. and Fryer, M.J. (1980). Density estimation and suicide risks in psychiatric treatment. *J. Roy. Statist. Soc. A*, **143**, 167–176.

Cormack, R.M. (1971). A review of classification (with Discussion). *J. Roy. Statist. Soc. A*, **134**, 321–367.

Courant, R. and Hilbert, D. (1953). *Methods of Mathematical Physics*, Volume II. New York: Interscience.

Cover, T.M. and Hart, P.E. (1967). Nearest neighbor pattern classification. *IEEE Trans. Inf. Thy.*, **IT-13**, 21–27.

Cox, D.R. (1966). Notes on the analysis of mixed frequency distributions. *Brit. J. Math Statist. Psychol.*, **19**, 39–47.

Cox, D.R. and Hinkley, D.V. (1974). *Theoretical Statistics*. London: Chapman and Hall.

Cox, D.R. and Oakes, D. (1984). *Analysis of Survival Data*. London: Chapman and Hall.

Deheuvels, P. (1977). Estimation nonparametrique de la densité par histogrammes generalisés. *Rev. Statist. Appl.*, **35**, 5–42.

Delecroix, M. (1983). *Histogrammes et Estimation de la Densité*. Que sais-je? no. 2055. Paris: Presses Universitaires de France.

Devroye, L. and Györfi, L. (1985). *Nonparametric Density Estimation: The L_1 View*. New York: Wiley.

Duin, R.P.W. (1976). On the choice of smoothing parameters for Parzen estimators of probability density functions. *IEEE Trans. Comput.*, **C-25**, 1175–1179.

Efron, B. (1981). Nonparametric estimates of standard error: the jackknife, the bootstrap and other methods. *Biometrika*, **68**, 589–599.

Efron, B. (1982). *The Jackknife, the Bootstrap and other Resampling Plans*. Philadelphia: SIAM.

Emery, J.L. and Carpenter, R.G. (1974). Pulmonary mast cells in infants and their relation to unexpected death in infancy. In Robinson, R.R. (ed.), *SIDS 1974 Proceedings of the Francis E. Camps International Symposium on Sudden and Unexpected Deaths in Infancy.* Toronto: Canadian Foundation for the Study of Infant Deaths. pp. 7–19.

Epanechnikov, V.A. (1969). Nonparametric estimation of a multidimensional probability density. *Theor. Probab. Appl.,* **14.**153–158.

Feller, W. (1966). *An Introduction to Probability Theory and its Applications,* Volume II. New York: Wiley.

Fisher, R.A. (1936). The use of multiple measurements in taxonomic problems. *Ann. Eugen.,* **7,** 179–188.

Fix, E. and Hodges, J.L. (1951). Discriminatory analysis, nonparametric estimation: consistency properties. *Report No. 4, Project no. 21–49–004,* USAF School of Aviation Medicine, Randolph Field, Texas.

Friedman, J.H., Baskett, F. and Shustek, L.J. (1975). An algorithm for finding nearest neighbors. *IEEE Trans. Comput.,* **24,** 1000–1006.

Friedman, J.H., Bentley, J.L. and Finkel, R.A. (1977). An algorithm for finding best matches in logarithmic expected time. *ACM Trans. on Math. Software,* **3,** 209–226.

Friedman, J.H. and Tukey, J.W. (1974). A projection pursuit algorithm for exploratory data analysis. *IEEE Trans. Comput.,* **C-23,** 881–889.

Fryer, M.J. (1976). Some errors associated with the nonparametric estimation of density functions. *J. Inst. Maths. Applics.,* **18,** 371–380.

Fryer, M.J. (1977). A review of some nonparametric methods of density estimation. *J. Inst. Maths. Applics.,* **20,** 335–354.

Fukunaga, K. (1972). *Introduction to Statistical Pattern Recognition.* New York: Academic Press.

Fukunaga, K. and Hostetler, L.D. (1975). The estimation of the gradient of a density function, with applications in pattern recognition. *IEEE Trans. Info. Thy.,* **IT-21,** 32–40.

Geisser, S. (1975). The predictive sample reuse method with applications. *J. Amer. Statist. Assoc.,* **70,** 320–328.

Ghorai, J. and Rubin, H. (1979). Computational procedure for maximum penalized likelihood estimate. *J. Statist. Comput. Simul.,* **10,** 65–78.

Good, I.J. and Gaskins, R.A. (1971). Nonparametric roughness penalties for probability densities. *Biometrika,* **58,** 255–277.

Good, I.J. and Gaskins, R.A. (1980). Density estimation and bump-hunting by the penalized likelihood method exemplified by scattering and meteorite data. *J. Amer. Statist. Assoc.,* **75,** 42–73.

Gordon, A.D. (1981). *Classification.* London: Chapman and Hall.

Habbema, J.D.F., Hermans, J. and van der Broek, K. (1974). A stepwise discrimination program using density estimation. In Bruckman, G. (ed.), *Compstat 1974.* Vienna: Physica Verlag, pp. 100–110.

Habbema, J.D.F., Hermans, J. and Remme, J. (1978). Variable kernel density estimation in discriminant analysis. *Compstat 1978, Proceedings in*

Computational Statistics. Vienna: Physica Verlag.

Hall, P. (1983). Large sample optimality of least squares cross-validation in density estimation, *Ann. Statist.*, **11**, 1156–1174.

Hand, D.J. (1981). *Discrimination and Classification.* Chichester: Wiley.

Hand, D.J. (1982). *Kernel Discriminant Analysis.* Chichester: Research Studies Press.

Hartigan, J.A. and Hartigan, P.M. (1985). The dip test of unimodality. *Ann. Statist.*, **13**, 70–84.

Hellman, M.E. (1970). The nearest neighbor classification rule with a reject option. *IEEE Trans. Sys. Sci. Cyb.*, **SSC-6**, 179–185.

Hermans, J., Habbema, J.D.F., Kasanmoentalib, T.K.D. and Raatgever, J.W. (1982). *Manual for the ALLOC80 discriminant analysis program.* Dept. of Medical Statistics, University of Leiden, The Netherlands.

Hodges, J.L. and Lehmann, E.L. (1956). The efficiency of some nonparametric competitors of the *t*-test. *Ann. Math. Statist.*, **27**, 324–335.

Huber, P.J. (1985). Projection pursuit. *Ann. Statist.*, **13**, 435–475.

Jones, M.C. (1983). The projection pursuit algorithm for exploratory data analysis. Ph.D. Thesis, University of Bath.

Jones, M.C. and Lotwick, H.W. (1983). On the errors involved in computing the empirical characteristic function. *J. Statist. Comput. Simul.*, **17**, 133–149.

Jones, M.C. and Lotwick, H.W. (1984). A remark on Algorithm AS 176. Kernel density estimation using the fast Fourier transform. Remark AS R50. *Appl. Statist.*, **33**, 120–122.

Jones, M.C. and Sibson, R. (1987). What is projection pursuit? *J. Roy. Statist. Soc. A*, **150**, 1–36.

Kendall, M.G. and Stuart, A. (1973). *The Advanced Theory of Statistics*, Volume 2, 3rd Edition. London: Griffin.

Kent, J.T. (1976). Contribution to the discussion of a paper by P.R. Freeman. *J. Roy. Statist. Soc. A*, **139**, 39–40.

Kittler, J. (1976). A locally sensitive method for cluster analysis. *Pattern Recognition*, **8**, 23–33.

Koontz, W.L.G., Narendra, P.M. and Fukunaga, K. (1976). A graph-theoretic approach to nonparametric cluster analysis. *IEEE Trans. Comput.*, **C-25**, 936–943.

Kreider, D.L., Kuller, R.G., Ostberg, D.R. and Perkins, F.W. (1966) *An Introduction to Linear Analysis.* Reading, Mass: Addison-Wesley.

Leonard, T. (1978). Density estimation, stochastic processes and prior information (with Discussion). *J. Roy. Statist. Soc. B*, **40**, 113–146.

Loftsgaarden, D.O. and Quesenberry, C.P. (1965). A nonparametric estimate of a multivariate density function. *Ann. Math. Statist.*, **36**, 1049–1051.

Lubischew, A.A. (1962). On the use of discriminant functions in taxonomy. *Biometrics*, **18**, 455–477.

Mack, Y.P. and Rosenblatt, M. (1979). Multivariate K-nearest neighbor density estimates. *J. Multivariate Anal.*, **9**, 1–15.

Mardia, K.V. (1972). *Statistics of Directional Data*. London: Academic Press.

Mardia, K.V., Kent, J.T. and Bibby, J.M. (1979). *Multivariate Analysis*. London: Academic Press.

Monro, D.M. (1976). Real discrete fast Fourier transform. Statistical Algorithm AS 97, *Appl. Statist.*, **25**, 166–172.

de Montricher, G.M., Tapia, R.A. and Thompson, J.R. (1975). Nonparametric maximum likelihood estimation of probability densities by penalty function methods. *Ann. Statist.*, **3**, 1329–1348.

Müller, H.G. (1984). Smooth optimum kernel estimators of densities, regression curves and modes. *Ann. Statist.*, **12**, 766–774.

Nadaraya, E.A. (1965). On nonparametric estimates of density functions and regression curves. *Theor. Probab. Appl.*, **10**, 186–190.

Parzen, E. (1962). On estimation of a probability density function and mode. *Ann. Math. Statist.*, **33**, 1065–1076.

Parzen, E. (1979). Nonparametric statistical data modeling. *J. Amer. Statist. Assoc.*, **74**, 105–131.

Prakasa Rao, B.L.S. (1983). *Nonparametric Functional Estimation*. New York: Academic Press.

Remme, J., Habbema, J.D.F. and Hermans, J. (1980). A simulative comparison of linear, quadratic and kernel discrimination. *J. Statist. Comput. Simul.*, **11**, 87–106.

Rice, J. (1975). Statistical methods of use in analysing sequences of earthquakes. *Geophys. J.R. Astr. Soc.*, **42**, 671–683.

Rice, J. and Rosenblatt, M. (1976). Estimation of the log survivor function and hazard function. *Sankhya Ser. A*, **38**, 60–78.

Ripley, B.D. (1983). Computer generation of random variables: a tutorial. *Int. Stat. Rev.*, **51**, 301–319.

Rosenblatt, M. (1956). Remarks on some nonparametric estimates of a density function. *Ann. Math. Statist.*, **27**, 832–837.

Rosenblatt, M. (1971). Curve estimates. *Ann. Math. Statist.*, **42**, 1815–1842.

Rosenblatt, M. (1979). Global measures of deviation for kernel and nearest neighbor density estimates. In Gasser, T. and Rosenblatt, M. (eds), *Smoothing Techniques for Curve Estimation*. Lecture Notes in Mathematics, 757. Berlin: Springer-Verlag, pp. 181–190.

Rudemo, M. (1982). Empirical choice of histograms and kernel density estimators. *Scand. J. Statist.*, **9**, 65–78.

Schucany, W.R. and Sommers, J.P. (1977). Improvement of kernel type density estimators. *J. Amer. Statist. Assoc.*, **72**, 420–423.

Schuster, E.F. and Gregory, C.G. (1981). On the nonconsistency of maximum likelihood nonparametric density estimators. In Eddy, W.F. (ed.), *Computer Science and Statistics : Proceedings of the 13th Symposium on the*

Interface. New York: Springer-Verlag, pp. 295–298.

Scott, D.W. (1979). On optimal and data-based histograms. *Biometrika*, **66**, 605–610.

Scott, D.W. (1982). Review of some results in bivariate density estimation. In Guseman, L.F. (ed.), *Proceedings of the NASA Workshop on Density Estimation and Function Smoothing March 1982.* Dept of Mathematics, Texas A & M University, College Station, Texas, pp. 165–194.

Scott, D.W. and Factor, L.E. (1981). Monte Carlo study of three data-based nonparametric density estimators. *J. Amer. Statist. Assoc.*, **76**, 9–15.

Scott, D.W., Gotto, A.M., Cole, J.S. and Gorry, G.A. (1978). Plasma lipids as collateral risk factors in coronary heart disease – a study of 371 males with chest pain. *J. Chronic Diseases*, **31**, 337–345.

Scott, D.W., Tapia, R.A. and Thompson, J.R. (1977). Kernel density estimation revisited. *Nonlinear Analysis*, **1**, 339–372.

Scott, D.W., Tapia, R.A. and Thompson, J.R. (1980). Nonparametric probability density estimation by discrete maximum penalized-likelihood criteria. *Ann. Statist.*, **8**, 820–832.

Scott, D.W. and Thompson, J.R. (1983). Probability density estimation in higher dimensions. In Gentle, J.E. (ed.), *Computer Science and Statistics: Proceedings of the Fifteenth Symposium on the Interface.* Amsterdam: North-Holland, pp. 173–179.

Sibson, R. and Thomson, G.D. (1981). A seamed quadratic element for contouring. *Comput. J.*, **24**, 378–382.

Silverman, B.W. (1978a). Choosing the window width when estimating a density. *Biometrika*, **65**, 1–11.

Silverman, B.W. (1978b). Weak and strong uniform consistency of the kernel estimate of a density function and its derivatives. *Ann. Statist.*, **6**, 177–184. (Addendum 1980, *Ann. Statist.*, **8**, 1175–1176.)

Silverman, B.W. (1978c). Density ratios, empirical likelihood and cot death. *Applied Statistics*, **27**, 26–33.

Silverman, B.W. (1980). Density estimation: Are theoretical results useful in practice? In Chakravarti, I.M. (ed.), *Asymptotic Theory of Statistical Tests and Estimation.* New York: Academic Press, pp. 179–203.

Silverman, B.W., (1981a). Density estimation for univariate and bivariate data. In Barnett, V. (ed.), *Interpreting Multivariate Data.* Chichester: Wiley, pp. 37–53.

Silverman, B.W. (1981b). Using kernel density estimates to investigate multimodality. *J. Roy. Statist. Soc. B*, **43**, 97–99.

Silverman, B.W. (1982a). Kernel density estimation using the fast Fourier transform. Statistical Algorithm AS 176. *Appl. Statist.*, **31**, 93–97.

Silverman, B.W. (1982b). On the estimation of a probability density function by the maximum penalized likelihood method. *Ann. Statist.*, **10**, 795–810.

Silverman, B.W. (1983). Some properties of a test for multimodality based on kernel density estimates. In Kingman, J.F.C. and Reuter, G.E.H. (eds), *Probability, Statistics and Analysis*. Cambridge: Cambridge University Press, pp. 248–259.

Silverman, B.W. (1984a). Spline smoothing: the equivalent variable kernel method. *Ann. Statist.*, **12**, 898–916.

Silverman, B.W. (1984b). A fast and efficient cross-validation method for smoothing parameter choice in spline regression. *J. Amer. Statist. Assoc.*, **79**, 584–589.

Silverman, B.W. (1985a). Some aspects of the spline smoothing approach to nonparametric regression curve fitting (with Discussion). *J. Roy. Statist. Soc. B*, **46**, 1–52.

Silverman, B.W. (1985b). Penalized maximum likelihood estimation. In Kotz, S. and Johnson, N.L. (eds), *Wiley Encyclopedia of Statistical Sciences*, Volume 6. New York: Wiley, pp. 664–667.

Stone, C.J. (1984). An asymptotically optimal window selection rule for kernel density estimates. *Ann. Statist.*, **12**, 1285–1297.

Stone, M. (1974). Cross-validatory choice and assessment of statistical predictions (with Discussion). *J. Roy. Statist. Soc. B*, **36**, 111–147.

Tapia, R.A. and Thompson, J.R. (1978). *Nonparametric Probability Density Estimation*. Baltimore: Johns Hopkins University Press.

Thomson, G.D. (1984). Automatic contouring by piecewise quadratic approximation. Ph.D. Thesis, University of Bath.

Titterington, D.M., Murray, G.D., Murray, L.S., Spiegelhalter, D.J., Skene, A.M., Habbema, J.D.F and Gelpke, G.J. (1981). Comparison of discrimination techniques applied to a complex data set of head injured patients. *J. Roy. Statist. Soc. A*, **144**, 145–174.

Tukey, P.A. and Tukey, J.W. (1981). Graphical display of data sets in 3 or more dimensions. In Barnett, V. (ed.), *Interpreting Multivariate Data*. Chichester: Wiley, pp. 189–275.

Watson, G.S. and Leadbetter, M.R. (1964). Hazard analysis I. *Biometrika*, **51**, 175–184.

Wegman, E.J. and Wright, I.W. (1983). Splines in statistics. *J. Amer. Statist. Assoc.*, **78**, 351–365.

Weisberg, S. (1980). *Applied Linear Regression*. New York: Wiley.

Wertz, W. (1978). *Statistical Density Estimation: A Survey*. Göttingen: Vandenhoeck and Ruprecht.

Wertz, W. and Schneider, B. (1979). Statistical density estimation: a bibliography. *Int. Stat. Rev.*, **47**, 155–175.

Whittle, P. (1958). On the smoothing of probability density functions. *J. Roy. Statist. Soc. B*, **20**, 334–343.

Woodroofe, M. (1970). On choosing a delta-sequence. *Ann. Math. Statist.*, **41**, 1665–1671.

Author index

Subject index